# Early Life on Earth

# Aims and Scope Topics in Geobiology Book Series

**Topics in Geobiology** series treats geobiology – the broad discipline that covers the history of life on Earth. The series aims for high quality, scholarly volumes of original research as well as broad reviews. Recent volumes have showcased a variety of organisms including cephalopods, corals, and rodents. They discuss the biology of these organisms-their ecology, phylogeny, and mode of life – and in addition, their fossil record – their distribution in time and space.
Other volumes are more theme based such as predator-prey relationships, skeletal mineralization, paleobiogeography, and approaches to high resolution stratigraphy, that cover a broad range of organisms. One theme that is at the heart of the series is the interplay between the history of life and the changing environment. This is treated in skeletal mineralization and how such skeletons record environmental signals and animal-sediment relationships in the marine environment.
The series editors also welcome any comments or suggestions for future volumes.

**Series Editors**
Neil H. Landman, landman@amnh.org
Peter Harries, harries@shell.cas.usf.edu

For other titles published in this series, go to
http://www.springer.com/series/6623

# Early Life on Earth

## A Practical Guide

David Wacey
Centre for Microscopy, Characterisation & Analysis and
School of Earth & Environment
University of Western Australia
35 Stirling Highway
Crawley, WA 6009
Australia

Springer

Dr. David Wacey
Centre for Microscopy, Characterisation & Analysis and
School of Earth & Environment
University of Western Australia
35 Stirling Highway
Crawley, WA 6009
Australia

ISBN 978-1-4020-9388-3 ISBN 978-1-4020-9389-0 (eBook)
DOI 10.1007/978-1-4020-9389-0

Library of Congress Control Number: 2008938718



*Cover illustration:*
**Main Image Caption**
Illustration of the DNA double helix containing key words associated with the search for early life on Earth. (There are instances where we have been unable to trace or contact the copyright holder. If notified the publisher will be pleased to rectify any errors or omissions at the earliest opportunity)

Small Figure 1 – Biological stromatolites or physical sedimentary artefacts? Intriguing structures within the ~3,400 Ma Strelley Pool chert of Western Australia.

Small Figure 2 – Microfossil or abiogenic artefact? One of the intensely debated microstructures from the ~3,460 Ma Apex chert of Western Australia.

Small Figure 3 – Strelley Pool in the Pilbara of Western Australia where putative evidence of life has been found in ~3,400 Ma sandstone and chert units.

Printed on acid-free paper

springer.com

# Acknowledgement

This book could not have been completed without the kind donation of photographs by Roger Buick, Gary Byerly, Lawrie Duck, Katja Etzel, Grant Ferris, Shoichi Kiyokawa, Stephen Moorbath, Nora Noffke, Ian Ogilvie, Bill Schopf, Andrew (Steelie) Steele, Ken Sugitani, Yuichiro Ueno, Martin Van Kranendonk, Maud Walsh, and the permission of numerous publishing houses. Thanks also go to the Geological Survey of Western Australia, Archean Biosphere Drilling Project, Shire of Roebourne Visitor Centre, Marble Bar Mining Office and the National Oceanic and Atmospheric Administration for provision of photographs and help with fieldwork. Special thanks go to Martin Brasier, Owen Green, Matt Kilburn, Nicola McLoughlin and Cris Stoakes, not only for numerous photographs but also for their company during fieldwork and stories around the campfire. Martin also kindly agreed to write the introduction to this book. Much of the laboratory work would not have been possible without the help of the staff of the Centre for Microscopy, Characterisation and Analysis at the University of Western Australia, the staff of the Geophysical Laboratory, Carnegie Institute, Washington, DC, and the staff of the Department of Earth Sciences at Oxford University. Finally, I would like to thank my family and friends for their continued support of my career.

# Contents

## Part II An Atlas of Claims for Early Archean Life

# Setting the Scene: Milestones in the Search for Early Life on Earth

**M. D. Brasier**

Some 150 years ago, Charles Darwin was greatly puzzled by a seeming absence of fossils in rocks older than the Cambrian period (Darwin, 1859). He drew attention to a veritable Lost World that we now know to have spanned more than 80% of Earth History. To put our modern evidence and thinking about Precambrian life into perspective, this introduction will reflect upon the development of three key ideas in this field: the Victorian *Eozoon* controversy, the ongoing stromatolite debate, and the recent Apex microfossil debate.

"How on Earth did life begin?" This is one of the noblest questions we can ask in science. But it took well over a century from 1859 to gain an understanding of life in the Precambrian – the world before the Cambrian explosion of animals. Why did an understanding take so long? Arguably it was because it was, and still remains, a very big and very difficult problem. Its study now involves the whole of the natural sciences. Progress has been a matter of slow attrition. For most of this time, for example, there has been no concept of the vast duration of Precambrian time, nor any evidence for a distinct biota.

As explored below, each generation has come up with its own favourite solution to this question – *whence cometh life*? – only to watch each one fall as the next generation of science and scientists has arrived on the scene. As such, this story provides us with a salutary tale of 'paradigm shifts' that have taken place about every 50 years or so. And, as I shall explore below, this process is ongoing and continuous. It is no surprise then, to find that the majority of uniformitarian interpretations for Precambrian fossil assemblages established over the last 50 years now appear highly questionable. That is, of course, exactly how it should be.

Palaeontologists and biologists had struggled to answer questions about the emergence of animal life long before Charles Darwin was to unveil his theory of evolution in 1859. Erasmus Darwin, his illustrious grandfather, wrote about the origins of life in his treatise 'Zoonomia' (Darwin, 1794). In so doing, he almost anticipated a modern definition of life: material that responds to stimuli, grows, reproduces inaccurately, and evolves by the transmission of these changes.

---

M. D. Brasier
Department of Earth Sciences, Parks Road, Oxford, OX1 3PR, UK

D. Wacey, *Early Life on Earth; A Practical Guide*,

By 1837, radical science was leading towards a concept of "progressive development". This referred to geological evidence then unfolding about the great history of life, from a lack of fossils in the 'Primary' rocks, to marine invertebrates in the Silurian, towards land vertebrates by the Jurassic, and ultimately to modern man, the whole chain arising from some ultimate cause. The great Oxford geologist Charles Lyell explained away this troubling phenomenon as the result of incompleteness in the early fossil record (Lyell, 1837). His strict adherence to the Principle of Uniformity was arguably a necessary step: an unwillingness to accept negative evidence; or determinism; and the need for a null hypothesis against which to test the claims of progressive development. Until we have evidence to the contrary, Lyell was saying, then we should assume that everything in the past worked in exactly the same way as we see now.

A wide but controversial airing in polite society was, however, given to the idea of evolution when Robert Chambers published his anonymous musings about progressive development and evolution (Chambers, 1844). But strong establishment reactions against his account, together with the barely known fossil record, still spoke against the suggestion that all of life shared a common ancestor. Anticipating other legitimate, or establishment, criticisms against his theory, Darwin was therefore obliged to speak cautiously about the absence of ancestors or obvious intermediates between the known animal groups: "*There is another difficulty which is much graver. I allude to the manner in which numbers of species of the same group, suddenly appear in the lowest known fossiliferous rocks. … I cannot doubt that all Silurian trilobites have descended from some one (form), which must have lived long before the Silurian age. … If my theory be true, it is indisputable that before the lowest Silurian stratum was deposited, long periods elapsed, as long, or probably far longer than, the whole interval from the Silurian age to the present day; and that during these vast, yet quite unknown, periods of time, the world swarmed with living creatures … the case at present must remain inexplicable; and may be truly urged as a valid argument against my views here entertained*" (Darwin, 1859).

Darwin's evolutionary theory of 1859 gave, of course, a revolutionary and coherent significance to the search for 'increasing organism complexity' through the rock record, as both Huxley's collected essays (see Huxley, 1894) and Haeckel's embryology (Haeckel, 1872) over the next decade show. But as we shall see, progress in the last half of the nineteenth century was mired in the famous *Eozoon* debate. A brief look at this debate is therefore rather instructive.

## 1 The *Eozoon* Debate and the 'Foraminosphere'

While Charles Darwin was polishing off the Origin of Species in 1859, Sir William Logan in Canada was contemplating what he took to be the world's oldest fossil. This consisted of thin layers of green serpentine alternating with calcite (Fig. 1a). It had been found on the banks of the River Ottawa to the west of Montreal, within a bed of marble some 500 m thick, intermixed with thick layers of banded gneiss and micaceous schist, now known to be some 1,100 million years old. Sir Charles

Lyell provides us with a description of the fossil as it seemed to him in 1865: "*It appears to have grown one layer over another, and to have formed reefs of limestone as do the living coral-building polyp animals. Parts of the original skeleton, consisting of carbonate of lime, are still preserved; while certain interspaces in the calcareous fossil have been filled up with serpentine and white augite*" (Lyell, 1865).

In 1864, this strange rock had been shown to Dr J.W. Dawson of Montreal – a one-time pupil of Lyell – who named it the 'Dawn Animal of Canada'. It is worth observing that Dawson was equally intrigued by the abundance of carbon – in the form of graphite – to be found in rocks with *Eozoon*. Not unreasonably for the time, he inferred that this pointed to some kind of vegetation long ago, though quite what kind of vegetation he could not say (Dawson, 1888). *Eozoon* was duly taken on world tour by Sir William Logan, to be displayed before the Geological Society in London. There, the eminent microscopist William B. Carpenter was struck by a seeming resemblance between ancient *Eozoon* (Fig. 1a) and some living foraminifera, such as *Discospirina* (Fig. 1b, which he called '*Orbitolites*') as well as some rock-forming types such as *Homotrema*. The white layers of marble were therefore regarded, by both Dawson and Carpenter, as the remains of skeletons that had grown, layer by layer, to contribute towards great reefs of limestone. These layers were then infilled by serpentine by some uncertain process.

Decoding of such 'metamorphic' rocks was to remain enigmatic – indeed it was largely guesswork – until about 1880. It was only then that the real advances came, following hard upon the heels of the petrographic microscope, the polariser, the analyser, the rotating stage and the diamond wheel for rock cutting. Thus it was, for nigh on 20 years after publication of 'the Origin of Species', that decoding the oldest crystalline rocks – and hence of *Eozoon* – was a difficult endeavour.

But with both Dawson and Carpenter enthralled by the resemblance between *Eozoon* and foraminiferids, both Lyell and Darwin were caught in a trap. Darwin was therefore moved to write, in the later editions of the Origin of Species: "*and the existence of the Eozoon in the Laurentian Formation of Canada is generally admitted. There are three great series of strata beneath the Silurian system in Canada, in the lowest of which Eozoon is found, Sir W. Logan states that their "united thickness may possibly far surpass that of all the succeeding rocks from the base of the Palaeozoic series to the present time. We are thus carried back to a period so remote, that the appearance of the so-called Primordial fauna (of Barrande) may by some be considered as a comparatively modern event." The Eozoon belongs to the most lowly organized of all classes of animals, but is highly organized for its class; it existed in countless numbers and, as Dr. Dawson has remarked, certainly preyed on other minute organic beings, which must have lived in great numbers. Thus the words which I wrote in 1859 about the existence of living beings long before the Cambrian period, and which are almost the same with those since used by Sir W. Logan, have proved true.*" (Darwin, 1871).

This concept of finding large and complex deep sea foraminifera in the oldest rocks needs to be placed in its proper historical context. In the 1840s, the deep sea world had not yet been explored. The geologist Edward Forbes had speculated that progressively more primitive forms would be found alive as deeper and deeper waters were sampled.

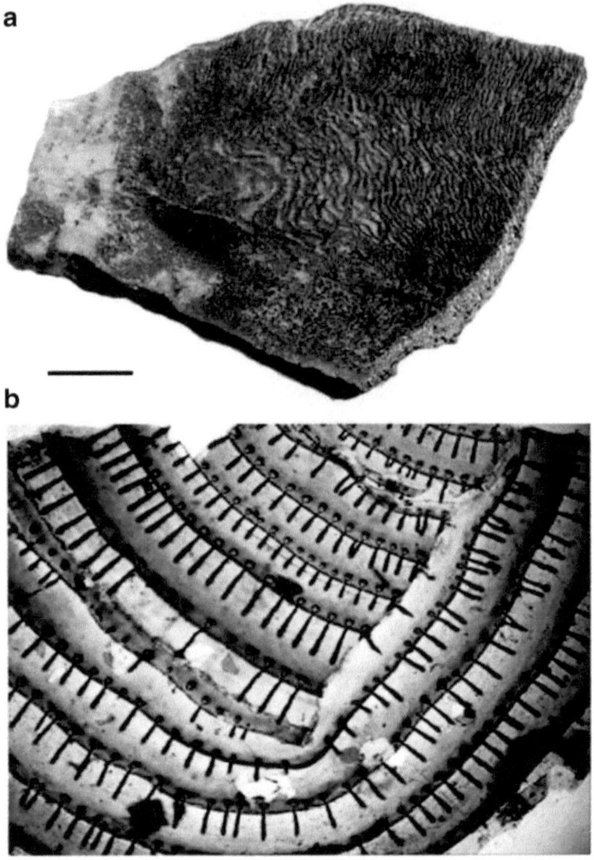


**Fig. 1** The Protozoan quest – foraminifera as the search image for the earliest life. (**a**) Hand specimen of *Eozoon canadense* (Carpenter, 1864), collected from the ~1,100 million year old ophicalcite of Cote-St-Pierre, Quebec, Canada. This was thought by Darwin, Dawson and Carpenter to provide the earliest evidence for life, and interpreted as a foraminiferid like the specimen below. (**b**) Transmitted light micrograph of the empty test of living foraminiferid *Discospirina*, imaged from the W.B. Carpenter collection held at Exeter University. Scale bar: 1 cm for (**a**); 100 μm for (**b**)

If so, deepest oceans would probably prove completely barren of life. That being so, the water column was thought to be a mirror of the history of life, with a Precambrian world in the deepest parts and a modern world in the shallows. But Forbes great idea was to receive a knock on the head in 1856, when British and American naval ships began to survey the deep Atlantic in readiness for the laying down of cables for the new electric telegraph. They found, of course, those vast carpets of tiny foraminiferal tests that we now call *Globigerina* Ooze. At the time, these globigerine tests were argued by Thomas Huxley – 'Darwin's Bulldog' – to havc lived on the seafloor and not in the water column as we now understand (Huxley, 1893–94). One of the earliest students of this chalky deposit was also one W.B. Carpenter.

As we can now appreciate, Carpenter and Dawson were to combine two distinct strands in their thinking about the early ancestors of life on Earth: that they will have been like those being found in the deep sea today – much as Forbes had suggested. And that that they will be like benthic foraminifera – much as Huxley believed. That is arguably why *Eozoon* fitted the search image – it was thought to be a pre-Cambrian benthic foraminiferid that hailed from the dawn of life itself.

But problems were soon to emerge. The first was the inferred but mistaken similarity between ancient *Eozoon* and complex living foraminifera such as *Discospirina*. This had led both Carpenter and Dawson to conclude that foraminifera had barely evolved since the 'Laurentian' (Proterozoic) period (see Darwin, 1871). But then a second set of observations sorted all this out – *Eozoon* was not a *fossil* at all. That was first shown by Irish geologists William Kind and Thomas Rowney in 1866, and later by the German microscopist Karl Mobius, in 1879. By the time of Darwin's death in 1882, the scientific community had condemned *Eozoon* to death as little more than a mineral growth, formed at great depth and high temperature. In Scotland, it was found next to major faults and intrusions. And in Italy it was seen coming out of a vent in Vesuvius (see Hofmann, 1971). The game was up and hard lessons had to be learned.

What strikes us first was how strange this scientific paradigm of the mid nineteenth century sounds to us now. The prediction at that time was that early life would somehow resemble those things now found living in the deep sea, namely benthic foraminifera. Not only that, but very complex foraminifera like those we can find today, such as *Homotrema*. But we now know that foraminifera such as *Homotrema* are highly adapted to a world fit for metazoans because they use sponge spicules to trap their prey; capture tiny zooplankton in the water column; and have calcium carbonate shells to stop them being eaten alive by fish. None of this was known back in 1866. This lack of understanding caused both Dawson and Carpenter to lead the world up the garden path. They believed that seemingly simple organisms like foraminifera had barely evolved at all – a view we can now see to be greatly in error.

An important lesson to be learned from *Eozoon* is this one: that we must not expect to find modern kinds of creature in the rock record before the Cambrian. The Earth before animals was like an alien planet.

## 2 The Cyanosphere, Phase 1

The place of *Eozoon* was quickly taken by that of the structures we now like to call 'stromatolites'.

Among the first to notice and wonder about the strange forms of stromatolitic growth was Charles Darwin, back in the 1830s, who included a figure (Fig. 2a) in his published 'Journal of Researches' and made the first ever pertinent observations on splash-zone stromatolites during the voyage of HMS Beagle: "*The rocks of St Paul appear from a distance of a brilliantly white colour. This is partly owing to a*

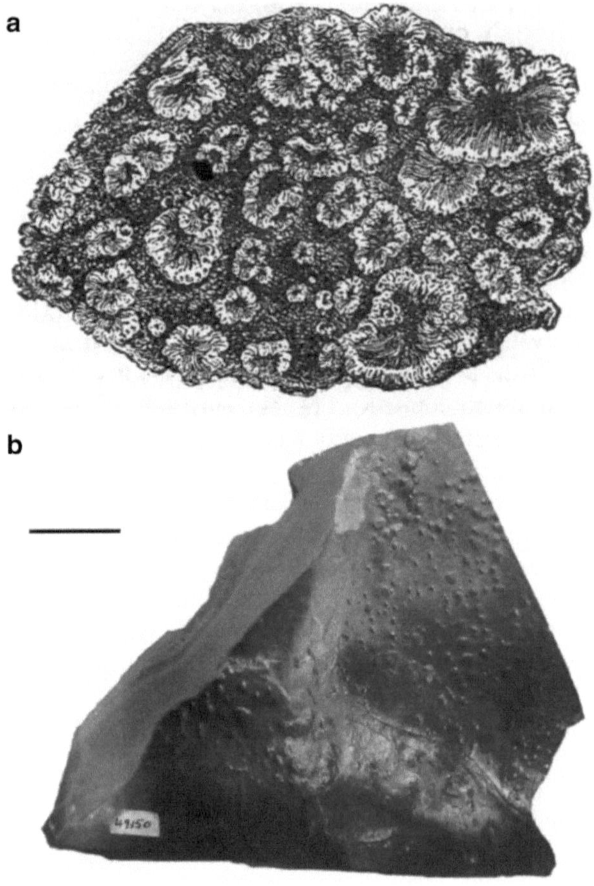


**Fig. 2** The Cyanobacterial quest – the discovery of self-organized sedimentary structures. (**a**) A woodcut used by Charles Darwin to illustrate vegetable-like growths of calcium phosphate found growing in the splash zones of oceanic islands; no scale was given. (**b**) One of several specimens of '*Arenicolites sparsus*' collected by J.W. Salter in 1856 from the Ediacaran Longmynd beds of England and illustrated by him the following year (Salter, 1857, pl. 5, Fig. 3). It is arguably among the first Precambrian fossils ever to be figured, named and described. This specimen is now regarded as a microbially-induced sedimentary structure. Scale bar is 5 cm

*coating of a hard glossy substance with a pearly lustre, which is intimately united to the surface of the rocks. This, when examined with a lens, is found to consist of numerous exceedingly thin layers, its total thickness being about the tenth of an inch. It contains much animal matter, and its origin, no doubt, is due to the action of rain or spray on the birds' dung. ... When we remember that lime, either as phosphate or carbonate, enters into the composition of hard parts, such as bones and shells, of all living animals, it is an interesting physiological fact to find substances harder than the enamel of teeth, and coloured surfaces as well polished as those of a fresh shells reformed through inorganic means from dead organic matter – mocking, also, in shape some of the lower vegetable productions."* (Darwin, 1839). We can only guess that he was thinking about comparisons with fungi and lichens.

The earliest microbially-mediated sediment to be figured from the Precambrian was discovered as early as 1855, by the great palaeontologist John Salter (see Fig. 2b) in the Ediacaran sediments of the Longmynd in England (Salter, 1856, 1857). This material was originally regarded by him as the markings of worm burrows of Cambrian age, but its Precambrian and microbial origin has now become well known. Interestingly, the specimen in Fig. 2b was indirectly referred to by Darwin (1859) while writing about the Origin of Species, as follows: "*Traces of life have been detected in the Longmynd beds beneath Barrande's so-called primordial [Cambrian] zone*" (Darwin, 1859).

From at least 1851, we also find discussions taking place about stromatolitic structures in the ~2,000 million year old Banded Iron Formations from the Great Lakes region of North America (Gunflint chert, see Fig. 3a). Such ancient rocks were then placed in the Azoic period because of their "*entire absence of organic*

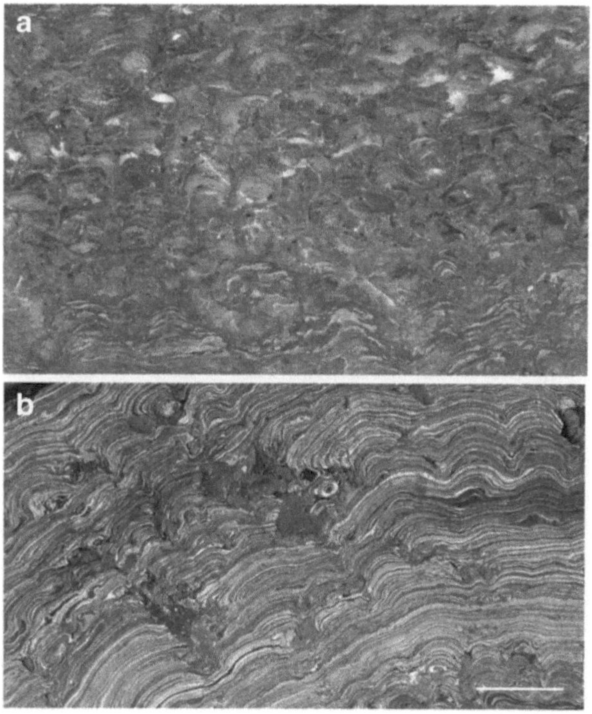


**Fig. 3** The Cyanobacterial quest – the discovery and decoding of self-organized sedimentary structures. (**a**) Hand specimen of a stromatolite from the ~2,000 million year old banded ironstones of the Gunflint chert, Minnesota. Similar structures were interpreted as igneous features by Foster and Whitney (1851) and as paradoxical sedimentary structures by Irving (1883). (**b**) Hand specimen once in the collection of Sir George Taylor, Director of the Royal Botanic Gardens at Kew. The specimen is made from layers of lead-based paint and was likely generated in a spray booth without any participation from biology. Note its similarity to 'a', including the non-isopachous laminae and the inter-columnar spaces filled with matrix. Scale bar is 2 cm for both images

*remains*" (e.g., Foster and Whitney, 1851). At that early time in geological thinking, the Gunflint chert was thought to have been laid down under the great residual heat left over from the fiery origins of the primordial Earth: "*The Azoic period having been one of long continued and violent mechanical action* (ibid., p. 67)." The concern, therefore, was whether the "*beautiful series of intricate convolutions of alternate bands of bright-red and steel-grey*" were really the result of sedimentary processes. "*The flexures are exceedingly intricate and bear no marks of having been the result of original stratification. There is no actual line of separation between the lighter and darker bands*" (ibid., p. 68). Their final suggestion was that they were crumpled by igneous phenomena and had "*risen up, in a plastic state from below*" to pour out onto the deep sea floor.

But the geological surveyor Roland Irving was to return to this question in his early monograph on the Archean of the north-western United States. By 1883, it was becoming clearer that the banded iron formations were some kind of paradoxical sedimentary deposit: "*the cherty and jaspery portions, frequently strongly charged with magnetite and other oxides of iron, present often peculiar irregularities and contortions in subordinate bedding, and also often a confused concretionary appearance, and even a brecciated appearance. All of these irregularities are very plainly subordinate to a simple bedding, corresponding entirely to that of the rest of the Animikie series. ... I anticipate that when we shall have completed our microscopic studies of them we shall get from them some light as to the origin of these confused and much discussed rocks. I may now merely say that ... all of these cherts and jasper schists are original, and not the result of a metamorphism upon ordinary sedimentary deposits, though manifestly they are not of eruptive origin, as has been maintained by some*" (Irving, 1883).

It was also at about this time that curious case of *Cryptozoon* began to emerge from the fossil record. This cabbage-like structure had been found spread across bedding planes in the upper Cambrian carbonates of New York State. First reported by James Hall in 1883, it was originally interpreted as some kind of calcareous algal growth, a view that was later developed by Charles Walcott and then by V.H.E. Kalkowsky (1908), who first introduced the concept of 'stromatolite' as follows: "*organogenic, laminated calcareous rock structures, the origins of which is clearly related to microscopic life, which itself must not be fossilized*" (translation in Krumbein, 1983). Stromatolites are so named from the Greek for 'flat stones'. But as can be seen, the definition of Kalkowsky was a genetic one, though direct observation of microbes was seemingly precluded by him. From this strange mismatch there has arisen much misunderstanding, not only about stromatolites but about the word 'stromatolite' itself (see Fig. 3b and McLoughlin et al., 2008).

The role of biology in stromatolite growth was widely ignored until Cambridge sedimentologist Maurice Black canoed across the tidal flats of Andros Island in the Bahamas (Black, 1933). He noticed that dense growths of cyanobacteria were forming an 'algal mat' that seemed to survive episodes of sediment deposition by growing upwards through the sediment.

Black's work went largely un-noticed for several decades. By 1954, the search for cyanobacteria-like microfossils in the Gunflint chert stromatolites was starting to bear some fruit in the form of fossilized cells, coccoidal colonies and filaments (Tyler and Barghoon, 1954; Barghoorn and Tyler, 1965). There followed an increasing number of descriptions of stromatolites, and of microfossils from stromatolitic cherts, in the succeeding decades (e.g., Walter, 1976; Krumbein, 1983). During this phase, it was widely assumed that stromatolites were largely formed by the trapping and binding activities of microbes such as living cyanobacteria. Such work was helping to promote and establish the paradigm that cyanobacteria are a search image for the earliest life, and that stromatolites are their constructions (see Schopf and Klein, 1992; Schopf, 1999).

This cyanobacterial vision of the early biosphere was to reach its zenith – or maybe its nadir – in the Viking Missions to Mars in 1976. During those expeditions, the chemical tests for early life were clearly designed to sniff out the evidence for cyanobacteria-like photosynthesis. Those early astrobiologists were looking for kinds of metabolisms that involved the uptake of carbon dioxide and the release of gaseous oxygen by means of photoautotrophic enzymes like chlorophyll. As is now well known, they found nothing to their liking.

The strangeness of distant planets and of early worlds really dates from this moment of realization back in 1976. But it has taken 3 more decades for the cyanobacterial paradigm – the cyanosphere – to crumble. The first nail in the coffin came with the discovery of stromatolite-like growth around the chimneys of deep sea 'black smokers'. Clearly, these ecosystems and the stromatolites they contain, cannot be photoautotrophic, let alone oxygenic. A major step forward was then taken by the MIT group of John Grotzinger and Dan Rothman (1996) who showed by experiment that stromatolitic growth, leading to branching and columnar forms, is the predictable outcome from some kinds of non-biological crystal growth, much like the growth of calcareous flowstone (cf. Fig. 4a). They suggested that such potentially non-biological stromatolites had isopachous laminae (Fig. 4b) whereas biologically mediated stromatolites tended to have non-isopachous laminae (Fig. 4c).

These conclusions were to have serious implications for the interpretation of stromatolites from the c. 3,400 million year old Strelley Pool Formation of Western Australia. First regarded as biological (Lowe, 1980), they were later rejected by him (Lowe, 1994), but further examples and arguments were resurrected by Hofmann et al. (1999) and then more recently by Allwood et al. (2006). My own studies (e.g., McLoughlin et al., 2008; Wacey et al., 2008), however, reveal that these structures typically show isopachous laminae, and form part of a spectrum that ranges from ripple-like corrugations of linear, through sinuous to linguoid and lunate forms, culminating in asymmetrically conical morphologies (Fig. 5). As such, they may be explained as accretionary flow-stone like bed forms formed under supersaturated conditions on the seafloor. As yet, there is no evidence in these stromatolites for the preservation of microbial fossils in the form of sheaths, filaments or cells.

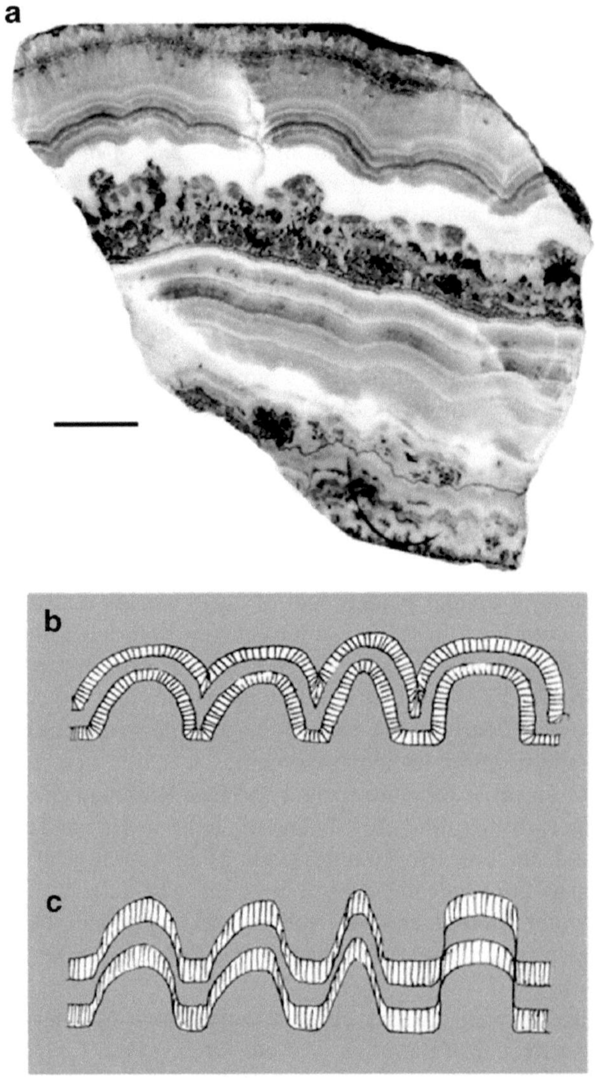


**Fig. 4** The Cyanobacterial quest – the decoding of self-organized sedimentary structures. (**a**) Hand specimen of banded agate showing stromatolite-like domes and bush-like dendrites, from an un-named deposit in Wyoming. Note the isopachous nature of the laminae, regarded by Groztinger and Rothman (1996) as potentially non-biological. Scale bar is 2 cm. (**b**) Sketch to show the nature of isopachous growth, typically found in agates, malachites and other hydrothermal mineral deposits. (**c**) Sketch to show the nature of non-isopachous laminae supposedly typical of stromatolites with a biological component

A further challenge to the paradigm of stromatolites as cyanobacterial markers – and even as biosedimentary structures – is now coming from experimental work at Oxford University, undertaken by Nicola McLoughlin and colleagues. They have managed

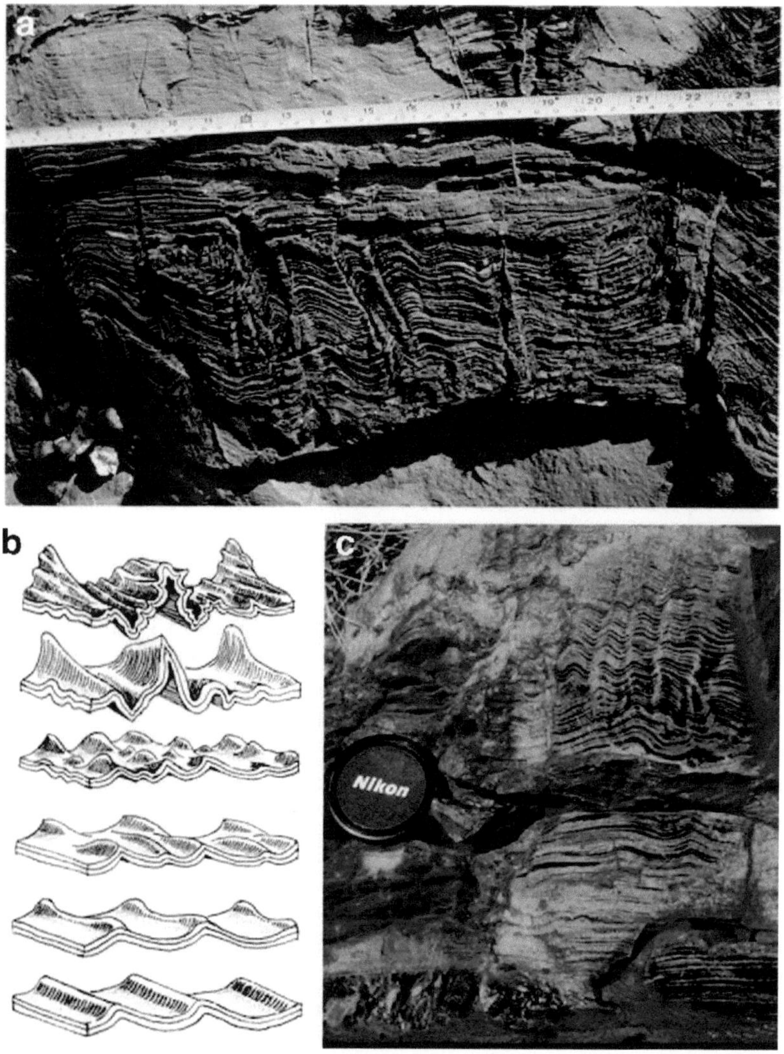


**Fig. 5** The Cyanobacterial quest – the decoding of self-organized sedimentary structures from the ~3,400 Ma Strelley Pool Formation. (**a**, **c**) Field photographs showing stacked pseudocolumns of linear, ripple-like features, commonly regarded as 'stromatolites'. (**b**) Reconstructions of the geometry of the so-called stromatolites from the Strelley Pool Formation, showing the continuous spectrum from linear ripples to oversteepened lunate ripples and pseudoconical structures. Scale is variable to fit observed morphologies

to grow complex digitate and inclined stromatolites with non-isopachous laminae (cf. Fig. 3b) under conditions that are relevant to the accretion of stromatolites (e.g., on the early land surface), and to the accretion of calcareous tufas and siliceous sinters (e.g., in the early oceans). These laboratory simulated stromatolites show some remarkable similarities with the famous Gunflint stromatolites (Fig. 3a).

From this and other studies, it emerges that stromatolite morphologies tend to accumulate along the edges of a deposition system where the supply of material is starting to fail – in what can be called the zone of complexity. At best, they may tell us something indirect about viscosity and the presence of extracellular mucilage. But clearly, they can also grow completely without the participation of biology. A plethora of studies have now shown that stromatolitic morphology should henceforth be regarded as a branch of physical sedimentology. That is to say, whereas ripples are the products of low viscosity accretion, stromatolites may be seen as the products of accretion under more viscous conditions. Clearly, this viscosity may take the form of either non-biologically- or biologically-induced crystal precipitates and gels. Both systems can produce self-organized structures of domes and columns under non-equilibrium conditions.

## 3 The Cyanosphere, Phase 2

The search for well-preserved ***cells*** in ancient rocks has a surprisingly long history. In the seventeenth century, Oxford microscopist Robert Hooke first observed the structures we call cells in the bark of the cork tree and the study of spores and pollen followed not long after. During the voyage on the Beagle in South America during the 1830s, Charles Darwin was moved to write about cherts from Chile: "*how surprising it is that every atom of the woody matter ... should have been removed and replaced by silex so perfectly, that each vessel and pore is preserved!*" (Darwin, 1839).

The earliest *bona fide* report of cellular preservation in Precambrian rocks seems to have been that made by Jephro Teall in the ~1,000 million year old Torridonian sedimentary phosphates, first reported in 1899 and soon after described and illustrated (see Peach et al., 1907). It was to be nearly another 50 years before comparable reports were to arrive from the ~2,000 million year old Gunflint chert (Tyler and Barghoon, 1954; Barghoorn and Tyler, 1965) and there were many followers to this work (see Schopf, 1999; Knoll, 2003).

This phase of research culminated, most famously, in the claim for a diverse suite of microfossils from the ~3,460 million year old Apex chert of Western Australia (Schopf, 1992, 1993, 1999). That work carried with it the implication that the origin of life likely took place about 4,000 million years ago on Earth. Life was then thought to have diversified rapidly by about 3,500 million years ago, culminating in the evolution of oxygen-releasing photosynthesis by cyanobacteria by that time (Schopf, 1993, 1999).

This concept of a Cyanosphere on the early Earth and Mars is now undergoing a critical scientific rethink. Brasier et al. (2002) began their challenge to this 'early Eden paradigm' by questioning the Earth's oldest supposed 'microfossil' assemblage, from the Apex chert. Schopf (1999) had inferred that eleven separate types of micro-organism were preserved in Apex cherts from Chinaman Creek, near Marble Bar in Western Australia. Of these, a number were compared with fossil and living cyanobacteria, with the major implication that oxygen was already being released into the atmosphere. The dubious nature of the Apex chert 'microfossil' assemblage came to light when the rock slices were examined in detail by Brasier and his co-workers for the first time since they were deposited, in the early 1990s, at the Natural History Museum in London. These 'microfossils', which include some of the smallest fossils ever named (*Archaeotrichion* is as small as 1/3,000th of a millimetre) were seen to grade into non-biological structures resulting from recrystallization of the rock fabric (see Fig. 6). Another structure, called *Eoleptonema apex*, seems to have grown down a post-depositional crack (Dr. A. Steele, 2008 personal communication). Such 'morphing' goes against one of the cardinal rules previously set up for the recognition of potentially biological structures (see pp. 44–50).

Given these concerns, Brasier et al. (2002) subjected the Schopf site and microstructures to new, high-resolution techniques, of the kind hitherto thought more appropriate for the study of Mars and potential Martian microfossils. Mapping was undertaken at a range of scales from kilometres to microns, and integrated with newly applied techniques for geochemistry and microfossil morphology. Together, these built up a completely new picture for the Apex chert 'microfossil' site, comprising hydrothermal fissure fillings rising towards the ocean floor through cracks in submarine basaltic lavas. These hot springs were being fed by a deep igneous heat source. The organic matter is very abundant in these dyke systems. Re-assessment of the Apex chert and its context revolves around eight major arguments (Brasier et al., 2002, 2005, 2006), summarized below.

1. The context for the 'microfossil' samples is not (as previously thought) a conglomerate formed on a beach or near the mouth of a river, but part of the feeder dyke/vein for a seafloor hydrothermal spring. This can be demonstrated by mapping, by fabrics and by geochemistry.
2. The putative 'microfossils' are not restricted to a distinctive class of clasts, often rounded (as was claimed). The structures actually occur in three successive generations of brecciated hydrothermal fissure fillings and glassy cements, while comparable structures occur in associated volcanic glass where temperatures must have reached 500°C, well above those viable for life. It could no longer be claimed, therefore, that all the fossils have simply fallen down the dyke from surface environments.
3. Associated structures that were once regarded as 'stromatolite-like clasts' were found to have fabrics that intergrade with laminated fissure fillings and are arguably laminites of non-biological origin.
4. The spatial arrangements of the 'microfossils' did not compare with that seen in the next oldest, diverse microfossil assemblage: that of the ~2,000 million year old Gunflint chert. There, filaments are wrapped around each other and clustered into layers that show clear behavioural orientation parallel to the laminae, whereas the 'microfossils' in the Apex chert show no coherent arrangement (e.g., Fig. 6) that might be thought consistent with biology.
5. The filaments are not all simple and unbranched (as previously thought). At least four of the holotypes have side branches, and all of them intergrade with adjacent branched structures. They form part of a morphological continuum that appears to

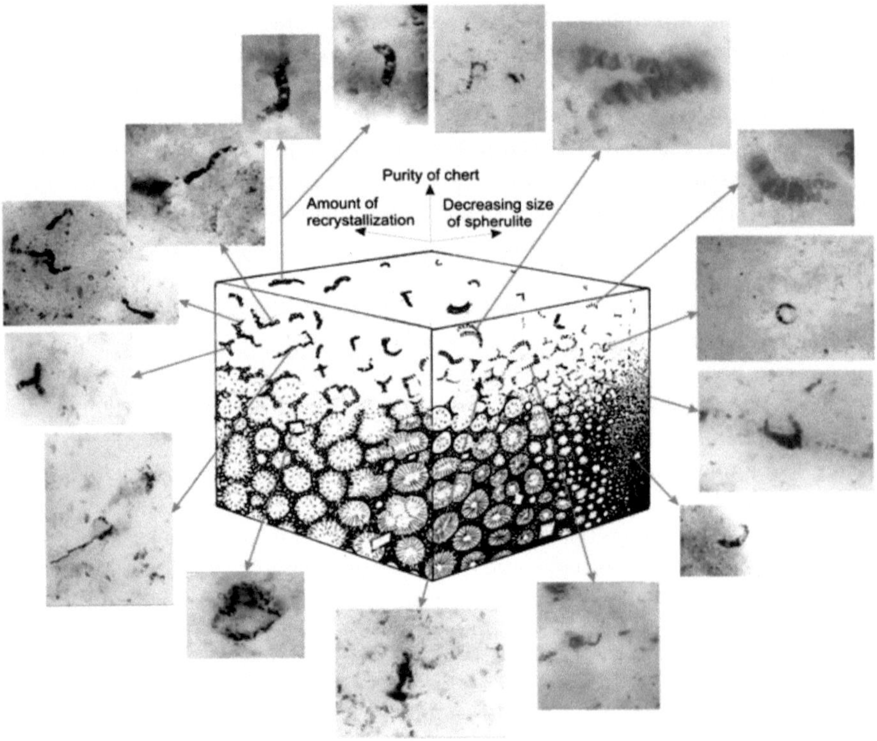


**Fig. 6** The Cyanobacterial quest – the decoding of self-organized hydrothermal structures from the ~3,460 Ma Apex chert. This three dimensional morphospace model (*centre block*) shows how this spectrum of microfossil-like structures (*outer images*) was most likely created entirely by physicochemical controls during recrystallization of the chert and the redistribution of carbonaceous material around spherulite and crystal margins. The key controls here were the relative purity of the chert (*vertical axis*), the degree of recrystallization of the fibrous chalcedony to equigranular microcrystalline chert (*left horizontal axis*), and the decreasing size of the spherulites (*right horizontal axis*). Arrows link theoretical with observed and reported microfossil-like artefacts having similar morphologies (see also pages 155–159). The scale is somewhat variable; the microfossil-like artefacts are between 10 and 100 μm in length

be due to the recrystallization of hydrothermal silica glass (see Fig. 6). As the silica recrystallized, it pushed carbonaceous impurities ahead of the radiating crystal fans because they could not be incorporated in the lattice. Such a process results in rounded sheets of carbon where the impurities are abundant, to branched, dendritic or simple arcuate filaments where the impurities become scarce.

6. The appearance of 'septa' (cell walls) and of 'bifurcated cells' (in the process of cell division), is also seen in the associated non-biological structures (of spherulitic and dendritic filaments). Both are reinterpreted as products of recrystallization, leading to interleaved quartz and carbonaceous matter.

7. The structures are indeed made of carbonaceous matter. But this carbon is no different from that seen in the associated non-biological artefacts with which they intergrade, nor does it differ significantly from that seen in disordered graphite or in carbonaceous meteorites. It is misleading, therefore, to infer a biological origin for this carbonaceous matter.
8. The ratio of the light stable isotope of carbon ($^{12}C$) to the heavier stable isotope ($^{13}C$) has been used as an indication of biological fractionation where this ratio (standardised against the Pee Dee Belemnite) is found to fall between c. –20 and –40 parts per thousand. But while carbon isotopes in this range are consistent with a biological origin for this carbon (possibly from hyper-thermophilic bacteria), it is important to note that a similar range of values can be produced by non-biological Fischer Tropsch-type synthesis (e.g., Holm and Charlou, 2001), also suspected in Archean dyke systems.

# 4 Implications

Several major conclusions can be drawn from these case histories. Clearly there is a need for a geological understanding of the context for early life on Earth or Mars, not merely a biological understanding. This means the acquirement of a thorough training in metamorphic, igneous and sedimentary petrology, and the study of context at a range of scales, including the premier disciplines of geological mapping and fabric mapping. Only in that way can scientists yet hope to form an opinion as to whether the context and burial history of the host rock is consistent with claims made for early life. The second conclusion is equally serious. It concerns facing up to a non-biological rather than biological origin for candidate morphological (or biogeochemical) signals from the early Earth or Mars. As we have seen with *Eozoon*, *Cryptozoon* and *Eoleptonema*, we can no longer afford to ignore the ways in which abiology can simulate earthly biology, let alone unearthly biology. The main problem here is that morphological complexity has for long been taken to be a keystone characteristic for the earliest fossils (e.g., Buick et al., 1981; Schopf, 1999). A basic understanding of self-organizing structures (SOS) and complexity is therefore an essential step if the early fossil record is to be correctly decoded. Unfortunately, complex structures do not require complex causes, as shown nearly a century ago by d'Arcy Thompson (1917). As we have seen with stromatolites, they can arise naturally in physico-chemical systems within the realms of 'chaotic' behaviour (Grotzinger and Rothman, 1996).

In Fig. 7, attention is therefore drawn to a range of physico-chemical gradients that can lead to the formation of macroscopic stromatolites (a) and ripples (b) as well as to microfossil-like structures generated by the growth of dendrites (e), 'coffee-ring' effects (f), polygonal crystal rims (g) and spherulites (h). In each of the systems shown, a move to the right of the diagram results in a loss of symmetry but a gain in morphological or temporal complexity towards the 'chaotic domain' (see Stewart and Golubitsky, 1992). This leads to a 'symmetry-breaking cascade',

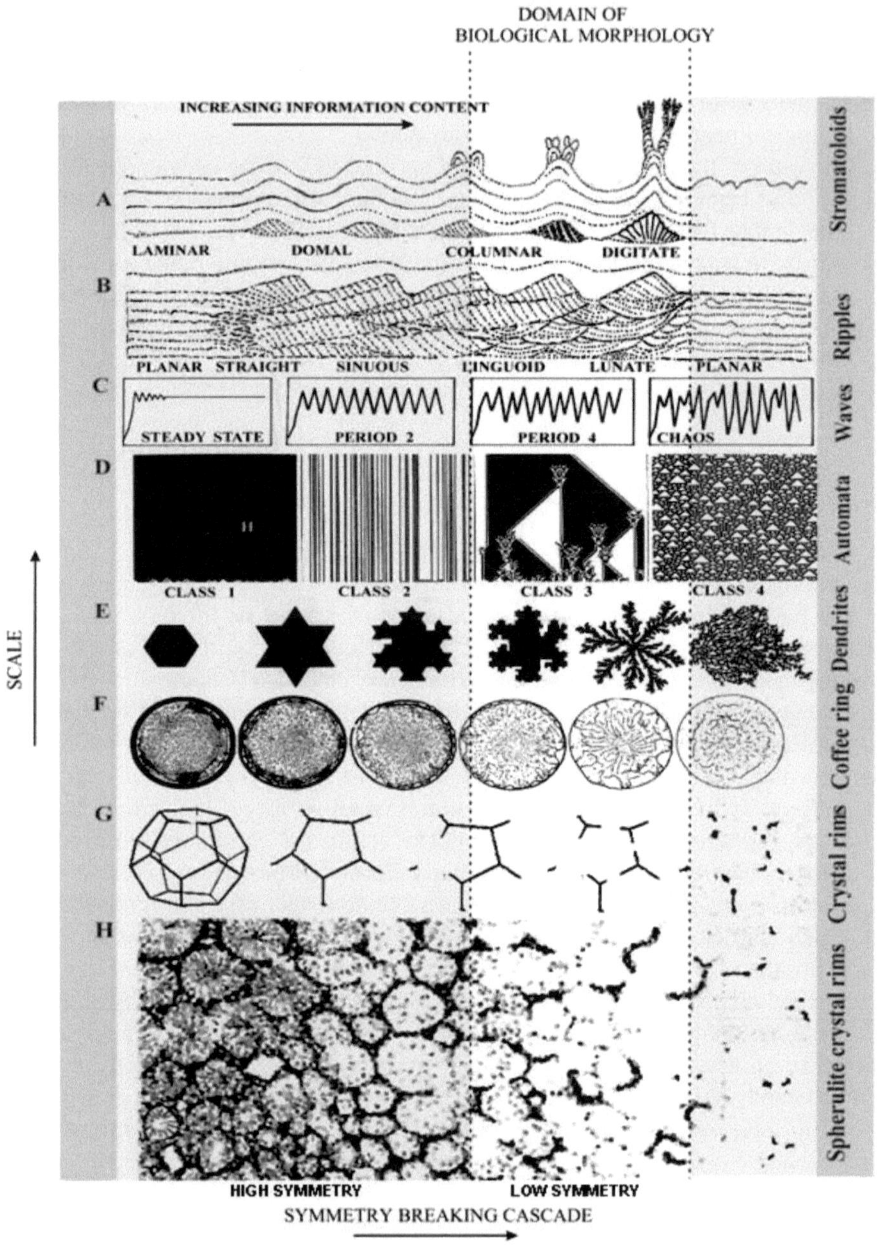


**Fig. 7** The range of self-organising structures (SOS) that can arise naturally in physico-chemical systems within the realms of chaotic behaviour. Symmetry is lost as one moves to the right but morphological complexity increases. In well preserved microfossil assemblages morphological variation of the fossil assemblages is usually less than co-occurring non-biological structures and so should occupy a more restricted domain ('domain of biological morphology') within the morphospace (From Brasier et al., 2006)

wherein the 'symmetry group' falls and the level of information rises. Symmetry-breaking is a particularly conspicuous phenomenon during the growth and recrystallization of spherulites, leading to natural assemblages of structures that can range from spheroidal (broadly rotational symmetry), to dendritic (reflectional to slide symmetry), to arcuate (no clear symmetry; Fig. 7h). Such symmetry-breaking cascades appear to arise when localised changes in the ionic concentrations of the constituent chemicals (e.g., iron oxide, carbon) fall below a critical threshold, so that the higher levels of symmetry became unstable. In this way, the margins of crystal growth can provide a rich harvest of pseudofossil structures, ranging from polygonal to dendritic to filamentous (e.g., snowflakes, moss agate, pyrolusite 'moss'; Fig. 7e–h) and from spherulitic/ botryoidal to dendritic to filamentous (e.g., hydrothermal cherts and jaspers; Fig. 7h). Such complex systems have also been simulated by computational experiments and digital automata (Fig. 7c–d), replicating the self-organization seen within stromatolites and dendrites (Grotzinger and Rothman, 1996; Wolfram, 2002). Brasier et al. (2006) have reviewed the problems of spheroids, filoids, septate filoids, stromatoloids, wisps and fluffs and the challenges that they present for decoding the earliest fossil record. Given such a challenge, it therefore seems wise to remain cautious and regard many Archean microfossils, stromatolites and carbon isotopic values (older than, say, c. 3,000 million years) as open to question until their origin by likely alternative, non-biological, processes has been falsified (the 'null hypothesis').

*The main aim of this book is therefore to draw attention to the remarkable number of highly interesting candidate structures that now await critical scrutiny by the next generation of explorers, in the quest to decode the earliest evidence for life on Earth.*

## Recommended Reading

Allwood, A. C., Walter, M. R., Kamber, B. S., Marshall, C. P., and Burch, I. W., 2006, Stromatolite reef from the Early Archaean era of Australia, *Nature* **441**: 714–718.

Barghoorn, E. S., and Tyler, S. A., 1965, Microorganisms from the Gunflint Chert, *Science* **147**: 563–577.

Black, M., 1933, Algal sediments of Andros Island, Bahamas, *Philosophical Transactions of the Royal Society B* **222**: 165–192.

Brasier, M. D., Green, O. R., Jephcoat, A. P., Kleppe, A. K., Van Kranendonk, M. J., Lindsay, J. F., Steele, A., and Grassineau N. V., 2002, Questioning the evidence for Earth's oldest fossils, *Nature* **416**: 76–81.

Brasier, M. D., Green, O. R., Lindsay, J. F., McLoughlin, N., Steele, A., and Stoakes, C., 2005, Critical testing of Earth's oldest putative fossil assemblage from the ~3.5 Ga Apex Chert, Chinaman Creek Western Australia, *Precambrian Research* **140**: 55–102.

Brasier, M. D., McLoughlin, N., and Wacey, D., 2006, A fresh look at the fossil evidence for early Archaean cellular life, *Philosophical Transactions of the Royal Society B* **361**: 887–902.

Buick, R., Dunlop, J. S. R., and Groves, D. I., 1981, Stromatolite recognition in ancient rocks: an appraisal of irregularly laminated structures in an early Archaean chert-barite unit from North Pole, Western Australia, *Alcheringa* **5**: 161–181.

Carpenter, W. B., 1864, On the structure and affinities of *Eozoon canadense*, *Proceedings of the Royal Society of London* **13**: 545–549.

Chambers, R., 1844, *Vestiges of the Natural History of Creation*. John Churchill, London.

Darwin, E., 1794, *Zoonomia; or The Laws of Organic Life*. Johnson, London.

Darwin, C., 1839, *Voyages of the Adventure and Beagle, Volume III – Journal and Remarks. 1832–1836*. Henry Colburn, London.

Darwin, C., 1859, *On the Origin of Species by Means of Natural Selection, or the Preservation of Favoured Races in the Struggle for Life (usually shortened to: On the Origin of Species)*. John Murray, London.

Darwin, C., 1871, *On the Origin of Species, 1871 edition*. John Murray, London.

Dawson, J. W., 1888, Specimens of *Eozoon canadense* and their geological and other relations, Montreal, Quebec. Peter Redpath Museum notes on specimens, McGill University, 106 p.

Foster, J. W., and Whitney, J. D., 1851, *Report on the Geology of the Lake Superior Land District*, pt. 2, The iron ranges. Sen. Documents: 32nd Cong., spec, sess., 3(4): 3–48.

Grotzinger, J. P., and Rothman, D. H., 1996, An abiotic model for stomatolite morphogenesis, *Nature* **383**: 423–425.

Haeckel, E., 1872, *Die Kalkschwamme. Eine Monographie*. Verlag von Georg Reimer, Berlin.

Hofmann, H. J., 1971, Precambrian fossils, pseudofossils and problematica in Canada, *Bulletin of the Geological Survey of Canada* **189**: 146 p.

Hofmann, H. J., Grey, K., Hickman, A. H., and Thorpe, R., 1999, Origin of 3.45Ga coniform stromatolites in Warawoona Group, Western Australia, *Bulletin of the Geological Society of America* **111**: 1256–1262.

Holm, N. G., and Charlou, J. L., 2001, Initial indicators of abiotic formation of hydrocarbons in the Rainbow ultramafic hydrothermal system, Mid-Atlantic Ridge, *Earth and Planetary Science Letters* **191**: 1–8.

Huxley, T. H., 1893–94, *Collected essays. 9 vols. Vol 1: Methods and Results; vol 2: Darwiniana; vol 3: Science and Education; vol 4: Science and Hebrew Tradition; vol 5: Science and Christian Tradition; vol 6: Hume, with Helps to the Study of Berkeley; vol 7: Man's Place in Nature; vol 8: Discourses Biological and Geological; vol 9: Evolution and Ethics, and Other Essays*. Macmillan, London.

Irving, R., 1883, *Copper bearing rocks of Lake Superior*. Monograph of the United States Geological Survey, 464 p.

Kalkowsky, V. H. E., 1908, Oolith und Stromatolith im Norddeutschen Buntsandstein, *Zeitschrift der Deutschen Geologischen Gesellschaft* **60**: 84–125.

Knoll, A. H., 2003, *Life on a Young Planet: The First Three Billion Years of Evolution on Earth*. Princeton University Press, Princeton, NJ, 277 p.

Krumbein, W. E., 1983, Stromatolites: the challenge of a term in space and time, *Precambrian Research* **20**: 493–531.

Lowe, D. R., 1980, Stromatolites 3,400-Myr old from the Archean of Western Australia, *Nature* **284**: 441–443.

Lowe, D. R., 1994, Abiological origin of described stromatolites older than 3.2 Ga, *Geology* **22**: 387–390.

Lyell, C., 1837, *Principles of Geology, 5th Edition*. John Murray, London.

Lyell, C., 1865, *Elements of Geology, 6th Edition*. John Murray, London.

McLoughlin, N., Wilson, L., and Brasier M. D., 2008, Growth of synthetic stromatolites and wrinkle structures in the absence of microbes: implications for the early fossil record. *Geobiology* **6**: 95–105.

Peach, B. N., Horne, J., Gunn, W., Clough, C. T., and Hinxman, L. W., 1907, *The Geological Structure of the Northwest Highlands of Scotland*, Memoirs of the Geological Survey of Great Britain.

Salter, J., 1856, On fossil remains of Cambrian rocks of the Longmynd and North Wales, *Quarterly Journal of the Geological Society of London* **12**: 246–251.

Salter, J., 1857, On annelide-burrows and surface markings from the Cambrian rocks of the Longmynd, *Quarterly Journal of the Geological Society of London* **13**: 199–206.

Schopf, J. W., 1992, The oldest fossils and what they mean, In: *Major Events in the History of Life* (ed. J. W. Schopf), John & Bartlett, Boston, pp 29–63.

Schopf, J. W., 1993, Microfossils of the Early Archaean Apex Chert: new evidence for the antiquity of life, *Science* **260**: 640–646.
Schopf, J. W., 1999, *The Cradle of Life*. Princeton University Press, Princeton, NJ, 367 p.
Schopf, J. W., and Klein, C. (Eds.), 1992, *The Proterozoic Biosphere: A Multidisciplinary Study*. Cambridge University Press, New York.
Stewart, I., and Golubitsky, M., 1992, *Fearful Symmetry. Is God a Geometer*. Penguin Science, London, 288 p.
Thompson, D'A. W., 1917, *On Growth and Form*. Cambridge University Press, Cambridge.
Tyler, S. A., and Barghoorn, E. S., 1954, Occurrence of preserved plants in pre-Cambrian rocks of the Canadian shield, *Science* **119**: 606–608.
Wacey, D., McLoughlin, N., Stoakes, C. A., Kilburn, M. R., Green, O. R., and Brasier, M. D., 2008, The ~3.4 Ga Strelley Pool Chert in the East Strelley greenstone belt – a field and petrographic guide. *Western Australia Geological Survey Record.*
Walter, M. R., 1976, *Stromatolites*. Elsevier, Amsterdam, 790 p.
Wolfram, S. 2002, *A New Kind of Science*. Wolfram Science, Champaign, IL, 1197pp.

# Part I
# Investigating Life in Early Archean Rocks

# Chapter 1
# What Can We Expect to Find in the Earliest Rock Record?

## 1.1 Introduction

The earliest preserved rock record, although fragmentary, provides us with unique evidence for testing models of when and where life first appeared on Earth. It is widely agreed that life emerged on our planet prior to 3,000 million years ago, but there currently exists no consensus as to the earliest fossil evidence of life on Earth.

In order for researchers to be able to work to a consistent baseline, we must first define "what is life?" This in itself is not a simple task (see Cleland and Chyba, 2002). For example, the current NASA definition of life as: "a system which is self-sustained by utilising external energy or nutrients owing to its internal process of component production and coupled to the medium via adaptive change that persist during the time history of the system" (Luisi, 1998) is both vague and awkward. For the purpose of this book a clearer, if more restrictive, definition is required which is tailored towards evidence that may feasibly be retrieved from the rock record. Namely that fossil life is "a complex structure that encodes evidence of biological behaviour and processing (e.g., growth, decay, and community tiering), and who's distribution and abundance is controlled by biologically significant variables such as light levels, temperature and nutrient gradients."

Such a definition allows signs of life to be manifested in three ways in the rock record:

- Firstly, as morphological remains (**body fossils**). Here the whole or part of an organism has been preserved within a rock. In most of the rock record only hard, mineralised parts of organisms (e.g., shells, bones) survive, but in exceptional circumstances, for example rapid burial or rapid mineralization, soft body parts may also be exquisitely preserved. Prior to 3,000 Ma, life was undoubtedly dominated by the most primitive of microorganisms lacking any hard parts. This means that we can only expect to find soft parts in the form of cells, sheaths and associated secreted polymeric substances preserved in the >3,000 Ma rock record. It follows that any search for body fossils must concentrate on the best preserved rock units.
- Secondly, as morphological **trace fossils**. These are non-body remains that in the broadest sense indicate the activity of an organism (e.g., dwellings, feeding tracks).

- Thirdly, as **chemical fossils**. These are traces of biological activity indicated by specific chemical signals left in rocks including: isotopic variations in, for example, carbon, sulphur, nitrogen or iron; distinctive ratios of elements; or molecular compounds that may be tied to a particular group of organisms.

## 1.2 Body Fossils

Organisms are divided into two large groups, the Prokaryota and the Eukaryota. The prime difference is that cells of the latter contain a nucleus. Eukaryotes make up all the multi-celled organisms such as plants and animals, and also some single celled organisms such as Amoeba. Prokaryotes are simpler organisms that lack not only a nucleus but also mitochondria and organised cell function. They are undoubtedly the simplest and most primitive of organisms, comprising the Bacteria and the Archaea, and predate unequivocal eukaryote cells in the fossil record by perhaps 2,000 Ma or more. There were clearly many crucial pre-cellular steps leading towards the origins of life and the first prokaryotes. These steps are likely to have included the development of an information transfer mechanism and the appearance of a cell wall to hold and concentrate the prebiotic chemicals. Locating these prebiotic processes in the rock record is inherently difficult and is in its infancy (an informative review of prebiotic chemistry can be found in Walde, 2005). For our purposes here, it is sufficient to be aware that any body fossils preserved in the earliest rock record should comprise these very simple (and very small) prokaryotic microbial cells.

The prokaryotic cell has a very simple internal structure (Fig. A1a). It does not have a membrane-bound nucleus and instead of having chromosomal DNA, its genetic information is in a central region called the nucleoid. Prokaryotic cells are very small, usually about 1–2 μm in diameter and up to about 10 μm long, and are of three main shapes: bacillus (rod shaped), coccus (spherical), and spirillum (spiral) (Fig. A1b). Prokaryotes have one distinct advantage over higher life forms in that they reproduce asexually. Instead of going through elaborate replication processes like eukaryotes, bacterial cells divide asexually by binary fission or budding. This type of reproduction allows for the duplication of great numbers of individuals in a rapid and simple way.

The body fossil record of the early Archean (>3,000 Ma) presents palaeobiologists with a preservational paradox. Microbial cells and sheaths, that are remarkably preserved from the late Archean onwards, have rarely been found in the early Archean. When they are found they are almost always controversial.

In seeking to explain this paradox we must first consider the preservational potential of the prokaryotic cell, the only likely morphological sign of life at this time. The preservation potential of each of the features of a cell is rather different. The cellular characteristic which tends to have the lowest chances of preservation is that of reproduction and its associated reproductive apparatus. This poor preservation may be because RNA and DNA molecules are intrinsically unstable and are readily degraded under heat and pressure. In eukaryotes, the nucleus may occasionally be

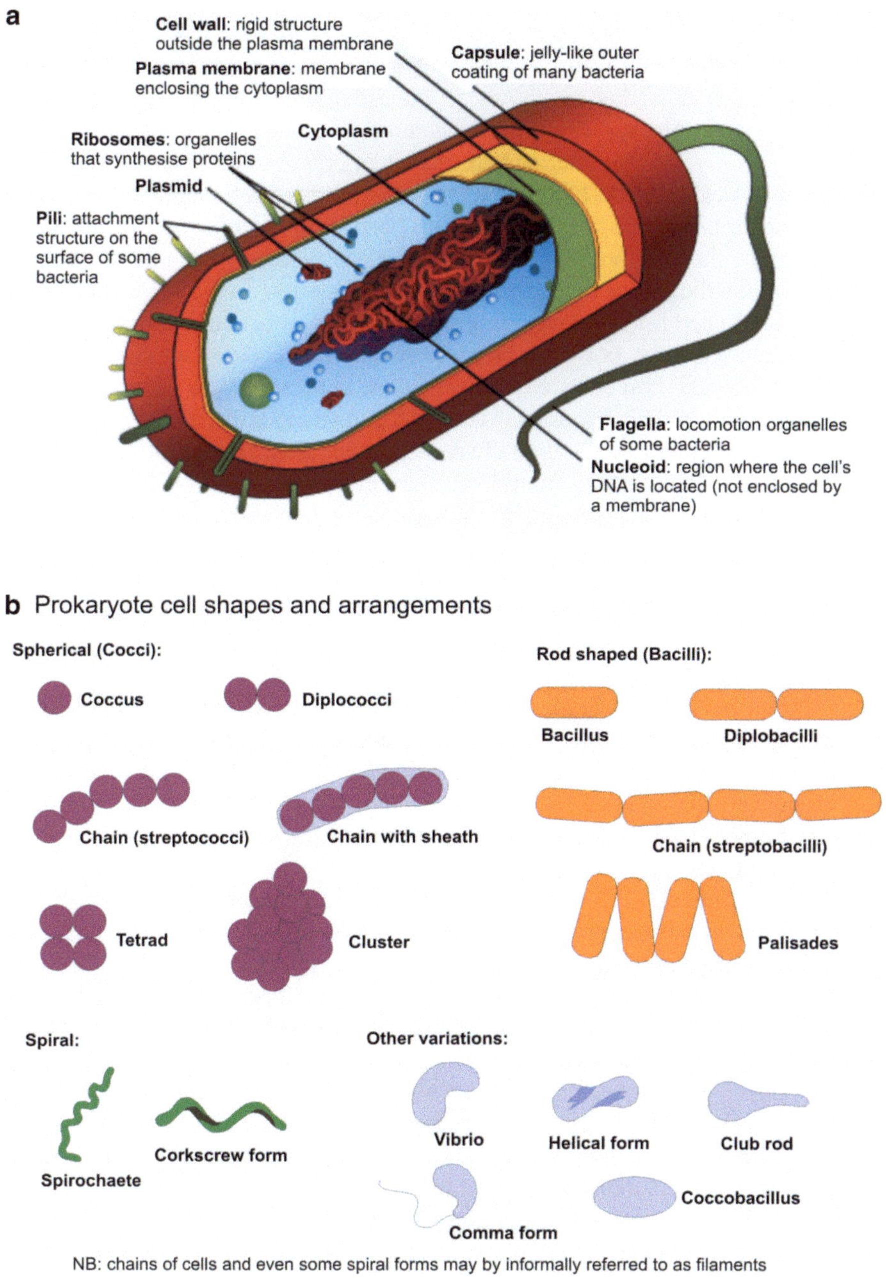


**Fig. A1** The prokaryote cell. (**a**) Structure and function of the cell. RNA messages are transcribed from DNA within the nucleoid; these messages are transformed into proteins in the ribosomes. Cellular metabolism is carried out by proteins and pigments within the cell's cytoplasm. (**b**) The variation in shape and arrangement of prokaryote cells. These are the key shapes to recognise when searching for cellular evidence in early Archean rocks preserved, but of course the nucleus is absent from prokaryotes. High resolution geochemical techniques (see for example Section 5.7) are now able to identify high relative concentrations of nitrogen within cell-like structures, however, and it may be that these can one day be shown to have derived from the degradation of reproductive materials. Preservation of the cell membrane in the early fossil record seems to have only a low to intermediate chance of preservation. The cell membrane of bacteria is largely made of series of phospholipids, often combined with fatty acids. The membrane acts as a permeability barrier for the transport of molecules into and out of the cell, but it is still weak and can be readily degraded, although less rapidly than that of the cell contents. This degradation is a natural consequence of the bacterial need for rapid recycling of raw materials (a chemical expression of the cell membrane may, however, be more readily preserved – see Section 5.12). The cell wall has the best chance of preservation. In bacteria, this is uniquely made of peptidoglycan, a polymer consisting of sugars and amino acids, and acts as a barrier against external forces. For the best chance of bacterial cell preservation, there is a requirement for rapid encapsulation of the cell wall within the preservational medium. The most stable media for this purpose appear to be silica (e.g., the ~1,900 Ma Gunflint chert, Fig. A4; Barghoorn and Tyler, 1965), iron sulphide (e.g., the ~3,200 Ma Sulphur Springs hydrothermal deposit; Rasmussen, 2000), or calcium phosphate (e.g., Doushantuo Formation; Xiao and Knoll, 1999). This preservation may arise from the metabolic processes of the cell itself, passive precipitation of an authigenic mineral matrix (entombing), or rapid burial protecting the cellular remains from oxidative degradation. For example, encrustations of the organism with a mineral precipitate may perhaps act as a UV shield, or serve to increase the proton motive force across the cell membrane, as with some iron oxidising bacteria (e.g., Chan et al., 2004); or conversely, the precipitate may be a metabolic by-product that would otherwise be detrimental to the metabolism of the cell.

In many prokaryotes, the cellular colony is often protected by extra-cellular polymeric substances (EPS) that have a relatively high chance of preservation. EPS has a good chance of preservation because it contains several active groups (e.g. carboxyl) on which mineral ions can nucleate to form mineralised biofilms. A good example is the extra-cellular cytoplasmic sheaths or envelopes found around the cells of cyanobacteria. The sheath is often preserved when the cells themselves have decomposed (e.g., 850 Ma Bitter Springs Formation; Oehler et al., 2006). The glutinous substances which comprise EPS can also have adhesive qualities trapping and binding sediment particles onto biofilms and bioaggregates, thereby leading to the formation of wrinkle structures and stromatolites. These structures have a reasonably good chance of preservation in the fossil record but, with the earliest examples, it is necessary to be careful not to confuse non-biological processes of formation, such as early diagenetic cements and hydrothermal silica gel precipitates. When preservation allows, stromatolitic structures constructed with the help of EPS should also show geochemical evidence for concentrations of biologically significant elements (e.g., C, N, P, S, plus trace elements that may commonly include Co, Fe, Ni, and Zn).

With all cellular and extra-cellular components, preservation is favoured by rapid occlusion of porosity and permeability because later migration of oxidising fluids will be in danger of wiping out the traces of cellular remains, introducing younger remains and/or altering the primary mineralogy of the rock. It has been demonstrated experimentally (Westall et al., 1995) that fossilization of bacteria can result in the complete replacement of the organic remains by a mineral crust or cast of the organism. Larger bacteria, such as cyanobacteria with thicker EPS sheaths, have more chance of retaining some organic material during fossilization. Although this type of fossilization preserves the size, shape and living habit of the bacteria, it nevertheless makes it more difficult to distinguish mineral artefacts from *bona fide* mineralized fossil remains. In these cases high resolution technology is required to identify any surviving traces of biology. Even the highest resolution technology is now being tested by the increasing number of reports of nano-bacteria. These can be as small as 30 nm and have been reported from hot spring carbonates and clays (e.g., Folk, 1999), and volcanic sands (Folk and Rasbury, 2002) where clustering and sediment boring appear to bear testament to their biological origin. Uwins et al. (1998) have even found evidence for DNA in modern nano-bacteria in an Australian sandstone. Nano-bacteria are particularly problematical; in rock samples they are usually mineralised and detection of organic matter is extremely difficult. Their simple morphology is non-diagnostic for biology, and they also resemble artefacts produced by SEM preparation. Convincing examples of nano-bacteria have yet to be demonstrated in Archean rocks.

## 1.3 Trace Fossils

Turning from cellular constituents to slightly larger and more visible signs of cellular activity, we come to microbial trace fossils. Trace fossils are morphological features preserved in the rock record that indicate the *activity* of some sort of life (which itself may not have been preserved). In the Phanerozoic rock record trace fossils are relatively easy to identify. Common examples include boreholes left in rocks by molluscs (Fig. A2a) and dinosaur footprints (Fig. A2b). Microbial trace fossils from the Archean, however, are very simple features and are very difficult to unambiguously identify. Those that have received the most attention are microbial borings. These are micron-sized cavities that have been created in rocks by the metabolic activities of microorganisms. Such microorganisms, that actively penetrate solid or semi-solid substrates, are termed euendoliths (Golubic et al., 1981). Because of their ability to actively bore into solid substrates, endolithic microbes are also one of the most common forms of modern contaminant in Archean rocks (e.g., Wacey et al., 2008a; Westall and Folk, 2003). Careful petrography and geochemistry is necessary to constrain the age of any endolithic microbes.

Microbial borings can preserve evidence for cellular morphology, microbial behaviour, ecology and metabolism in their selection and modification of rock substrates. They are well known from silicified carbonate sediments younger than about 1,600 Ma (e.g., Zhang and Golubic, 1987; Campbell, 1982) and have been widely reported from the glassy margins of modern pillow basalts (e.g., Fisk et al., 1998).

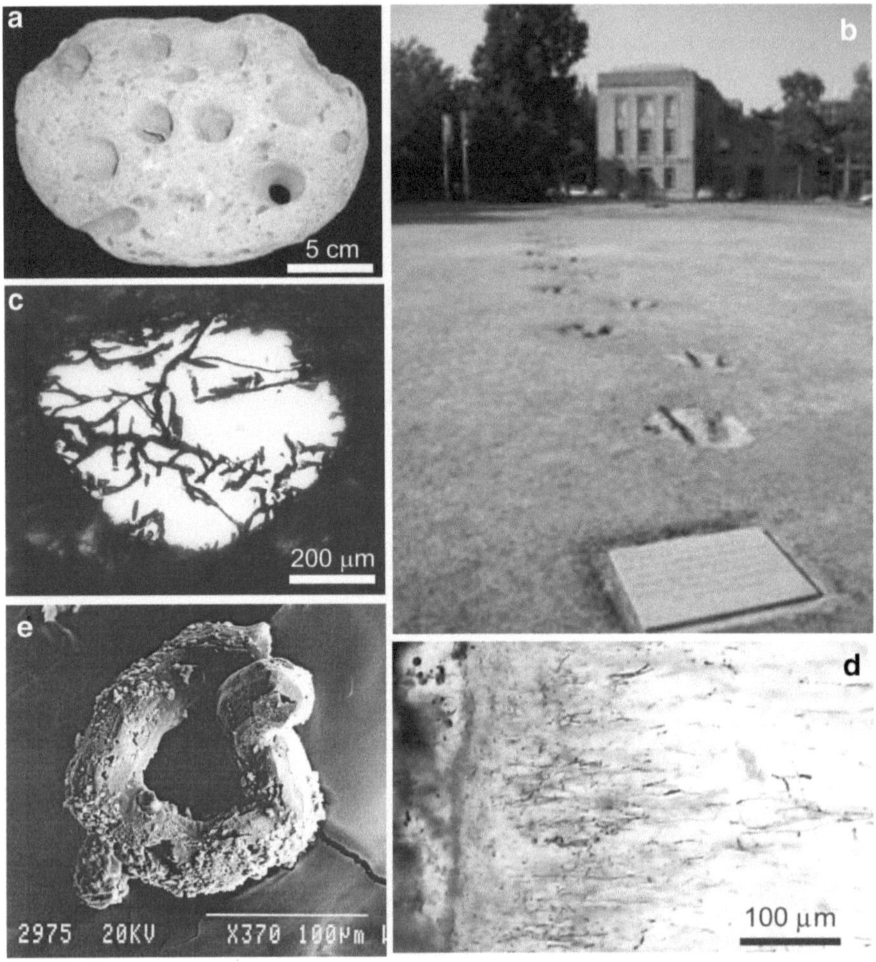


**Fig. A2** Examples of trace fossils from the rock record. (**a**) Boreholes in chalk made by the recent mollusc *Pholas*. (**b**) Three toed footprints made by the dinosaur *Megalosaurus* (Casts of the originals outside Department of Earth Sciences, Oxford University). (**c**) Cross section through a >3,400 Ma pyrite grain showing putative microbial borings. (**d**) Microbial borings preserved as linear arrays of fluid inclusions within a ~300 Ma brachiopod shell (From Buijs et al., 2004, reproduced with the permission of The Canadian Mineralogist). (**e**) Phosphate-filled ambient inclusion trail isolated from the lower Cambrian Soltanieh Formation of Iran, exhibiting typical angular termination and longitudinal striations (Image courtesy of Ian Ogilvie)

Importantly, putative microbial borings have recently been described from the margins of ~3,350–3,500 Ma pillow lavas from both South Africa and Australia (Furnes et al., 2004; Banerjee et al., 2006; Figs. B26, B27 and B63), and from >3,400 Ma metallic sandstone grains in the Pilbara of Western Australia (Figs. A2c and B43).

Preservation of microbial borings in the rock record usually requires early infilling with a stable mineral phase, often silica, prior to sediment compaction. However, it is also possible that microbial borings can be preserved without mineral infill in the form of fluid inclusion trails. Buijs et al. (2004) showed that fluid and gas filled endolithic microbial borings in ~300 Ma carbonates could be sealed by cementation and preserved as linear, curvi-linear and branching fluid inclusions that cut across crystal boundaries. Although these fluid inclusion trails may be broken down in patches by diagenetic modification, the presence of microbial borings can still be attested to by the preservation of linear arrays of inclusions (Fig. A2d). This may be the best search method for endolithic microbial borings in Archean carbonates, from which they are as yet unknown. In modern sandstones of the Ross Desert in Antarctica, crypto-endolithic bacteria inhabit pore spaces in the upper few millimetres of the sandstone (Friedmann and Weed, 1987); traces of this bacterial activity are preserved by a characteristic iron leaching patterns in the sandstone and irregularly shaped exfoliation flakes. Unfortunately, it is unlikely that these characteristic weathering patterns would be preserved in Archean rocks with such long and complex histories.

Ambient inclusion trails (AITs; Fig. A2e) are enigmatic microtubular structures which have in the past been confused with both microfossils and endolithic microbial borings. However, they can be distinguished by the following features: (1) presence of a mineral crystal (e.g., a metal sulphide or oxide) at one end of a microtube, of equivalent diameter to the tube, which may be pseudomorphed by later minerals (e.g., silica, metallic oxide, or phosphate as in Fig. A2e); (2) longitudinal striations on the microtube created by the angular edges of the propelled mineral crystal (may however be obscured by later mineral infill); (3) curved or twisted paths, particularly towards their ends as impedance of the host lithology affects movement; (4) tendency of microtubes to crosscut or form branches of a different diameter (i.e., where the propelled mineral becomes fragmented or a second crystal is intercepted), and to make sharp turns; (5) the microtube will likely have a polygonal cross sectional profile that matches the geometry of the propelled crystal.

Initially AITs were thought to be a completely inorganic feature, driven purely by forces of mineral cystallisation (Tyler and Barghoorn, 1963). Subsequently, however, a theory was advanced that AITs formed through the degassing of decomposing biological material trapped in an impermeable chert host, during burial and/or metamorphism (Knoll and Barghoorn, 1974). Recent high resolution geochemical studies of AITs from Western Australia (Wacey et al., 2008b) appear to support a biological component in their formation, although the exact formation mechanism remains elusive. Despite the fact that they are clearly different to either microfossils or biological microborings, AITs appear to hold potential for studies of early life.

## 1.4 Chemical Fossils

The metabolic processes and products of the earliest cells arguably have the highest chance of preservation in the Archean rock record. Although these processes may have little morphological expression, they inevitably modify the chemistry in and around the site of life. It is these chemical signatures that have the greatest chance of being preserved.

In order to track down the most likely chemical fossils preserved in the early rock record, we need to have an understanding of the range of metabolisms that prokaryotes are known to employ, and the range of environmental conditions on the early Earth that would have dictated which metabolisms were viable.

The early Earth is best viewed as an alien planet. The atmosphere, hydrosphere and lithosphere were very different to those we know today and were, in turn, imposing numerous constraints on biosphere evolution. The following conditions are thought by many to have existed at the surface of the Earth in the early Archean:

- Solar luminosity some 20% lower than now (Sagan and Mullen, 1972). If the Earth's atmospheric composition was similar to today, this 'faint young sun' would have led to widespread glaciations. Since there is no evidence of such glaciations in the rock record, and instead widespread evidence for liquid water, it has been suggested that the atmosphere contained large amounts of greenhouse gases (see below).
- An atmosphere of reducing gases that largely lacked oxygen (e.g., Kasting and Catling, 2003; Lowe and Tice, 2004). The implication of this is that the earliest microbes employed largely anaerobic metabolisms.
- No ozone layer to protect life from ultraviolet light (e.g., Konhauser et al., 2001). This would require the first microbes to have some shield to the higher UV flux. This could have been achieved by secretion of extra amounts of UV shielding polymers, or by maintaining an endolithic (within rocks) lifestyle.
- Much higher rates of solar and cosmic rays (e.g., Delsemme, 1998).
- High rates of meteoritic bombardment, with many over 10 km in diameter (e.g., Byerly et al., 2002; Moorbath, 2005). This 'late heavy bombardment' which lasted until about 3,850 Ma likely led to frequent vaporisation of the oceans and probably sterilised Earth's surface on a number of occasions. On the positive side, these same meteorites may have delivered exotic organic molecules (or even the first microbes) to the early Earth (Mileikowsky et al., 2000).
- A hot young crust, with higher rates of heat flux and hotter oceans (e.g., Knoll, 2003). A 'hydrothermal' ocean with temperatures of around 70°C, as proposed by Knauth and Lowe (2003), would have inhibited many microbial metabolisms including oxygenic photosynthesis. The temperature and chemistry of the Archean oceans is, however, highly debated. A pH of greater than 7 has been proposed (the so called 'soda ocean'; Kempe and Degens, 1985) although this is controversial. Less controversial are suggestions of oceanic silica saturation with extensive primary chert precipitation (Siever, 1992), and low calcium concentrations inhibiting $CaCO_3$ precipitation.
- The predominance of oceanic crust over granitic crust (Lowe, 1994). Mafic and ultramafic rocks are an excellent source of catalysts such as Ni and Co used by many biological enzymes. Thus, the highly metaliferous crust of the early Earth, when combined with enormous outflows of energy emanating from hydrothermal and volcanic systems, is likely to have played a significant role in both the genesis and sustenance of the earliest forms of life.

- A lack of extensive, modern style subduction zones and crustal recycling and a lack of large continental landmasses resulting in a restricted range of ecological niches (McCall, 2003; Van Kranendonk et al., 2004).

Following from this discussion, a logical conclusion may be that the first microbes were heat loving, flourishing around hydrothermal vents, utilising simple inorganic compounds of sulphur and/or nitrogen. Molecular phylogenetics also suggests that the first life form was probably a (hyper)thermophilic chemoautotroph (see Table 5.1 for definition). However, one of the great challenges for the current (and future) generations of scientists in this field is to better constrain the *types* of metabolisms operating on the early Earth. No matter what metabolism the first life utilized, it would in turn have initiated heterotrophic metabolic pathways (see Table 5.1) and multi-component ecosystems, as well as bio-weathering that could potentially modify the Earth's surface.

The most common examples of chemical fossils discovered thus far are metabolic fractionations in favour of the light isotopes of carbon and sulphur (see Sections 5.13 and 5.14), plus occasional remnants of cell membrane lipids preserved as hydrocarbons known as molecular fossils (Brocks et al., 1999; see Section 5.12). Other possible indicators of cellular metabolic processes may involve fractionation of the isotopes of nitrogen and iron, the highly localized storage of biologically significant, or even bio-limiting elements such as nitrogen and phosphorus, as well as trace elements important to biological enzymes (e.g., Ni, Co, Zn; see Williams and Frausto da Silva, 1996). To verify such biosignatures in the rock record, however, it is necessary to be able to discount similar fractionations or chemical enrichments arising from non-biological processes. Examples of plausible non-biological processes are so-called Fischer-Tropsch type reactions for the fractionation of carbon isotopes (see Fig. A35; Sherwood-Lollar et al., 2002; Horita and Berndt, 1999; McCollom and Seewald, 2006), or hydrothermal and photochemical fractionations of sulphur isotopes (see Section 5.14; Grassineau et al., 2001).

## Recommended Reading


Banerjee, N. R., Furnes, H., Muehlenbachs, K., Staudigel, H., and de Wit, M., 2006, Preservation of ~3.4–3.5 Ga microbial biomarkers in pillow lavas and hyaloclastites from the Barberton Greenstone Belt, South Africa, *Earth and Planetary Science Letters* 241: 707–722.

Barghoorn, E. S., and Tyler, S. A., 1965, Microorganisms from the Gunflint Chert, *Science* **147**: 563–577.

Brocks, J. J., Logan, G. A., Buick, R., and Summons, R. E., 1999, Archean molecular fossils and the early rise of eukaryotes, *Science* **285**: 1033–1036.

Buijs, G. J. A., Goldstein, R. H., Hasiotis, S. T., and Roberts, J. A., 2004, Preservation of microborings as fluid inclusions, *The Canadian Mineralogist* **42**: 1563–1581.

Byerly, G. R., Lowe, D. R., Wooden, J. L., and Xiaogang, X., 2002, An Archean impact layer from the Pilbara and Kaapvaal cratons, *Science* **297**: 1325–1327.

Campbell, S. E., 1982, Precambrian endoliths discovered, *Nature* **299**: 429–431.

Chan, C. S., De Stasio, G., Welch, S. A., Girasole, M., Frazer, B. H., Nesterova, M. V., Fakra, S., and Banfield, J. F., 2004, Microbial polysaccharides template assembly of nanocrystal fibres, *Science* **303**: 1656–1658.

Cleland, C. E., and Chyba, F., 2002, Defining 'Life', *Origins of Life and Evolution of the Biosphere* **32**: 387–393.

Delsemme, A. H., 1998, Cosmic origin of the biosphere. In: Brock, A. (Ed.) *The Molecular Origins of Life: Assembling the Pieces of the Puzzle*, Cambridge University Press, Cambridge, UK, pp. 100–118.

Fisk, M. R., Giovannoni, S. J., and Thorseth, I. H., 1998, The extent of microbial life in the volcanic crust of the ocean basins, *Science* **281**: 978–979.

Folk, R. L., 1999, Nannobacteria and the precipitation of carbonates in unusual sedimentary environments, *Geology* **126**: 47–55.

Folk, R. L., and Rasbury, E. T., 2002, Nanometer-scale spheroids on sands, Vulcano, Sicily: possible nanobacterial alteration, *Terra Nova* **14**: 469–475.

Friedmann, E. I., and Weed, R., 1987, Microbial trace-fossil formation, biogenous, and abiotic weathering in the Antarctic cold desert, *Science* **236**: 703–705.

Furnes, H., Banerjee, N. R., Muehlenbachs, K., Staudigel, H., and de Wit, M., 2004, Early Life recorded in Archean pillow lavas, *Science* **304**: 578–581.

Golubic, S., Friedmann, I., and Schneider, J., 1981, The lithobiontic ecological niche, with special reference to microorganisms, *Journal of Sedimentary Petrology* **51**: 475–478.

Grassineau, N. V., Nisbet, E. G., Bickle, M. J., Fowler, C. M. R., Lowry, D., Mattey, D. P., Abell, P., and Martin, A., 2001, Antiquity of the biological sulphur cycle: evidence from sulphur and carbon isotopes in 2700 million-year old rock of the Belingwe Belt, Zimbabwe, *Proceedings of The Royal Society of London B* **268**: 113–119.

Horita, J., and Berndt, M. E., 1999, Abiogenic methane formation and isotopic fractionation under hydrothermal conditions, *Science* **285**: 1055–1057.

Kasting, J. F., and Catling, D. C., 2003, Evolution of a habitable planet, *Annual Review Astronomy Astrophysics* **41**: 429–463.

Kempe, S., and Degens, E. T., 1985, An early soda ocean? *Chemical Geology* **53**: 95–108.

Knauth, L. P., and Lowe, D. R., 2003, High Archean climatic temperature inferred from oxygen isotope geochemistry of cherts in the 3.5 Ga Swaziland Supergroup, South Africa, *Geological Society of America Bulletin* **115**: 566–580.

Knoll, A. H., 2003, *Life on a Young Planet: The First Three Billion Years of Evolution on Earth*, Princeton University Press, Princeton, NJ, 277 p.

Knoll, A. H., and Barghoorn, E. S., 1974, Ambient pyrite in precambrian chert: new evidence and a theory, *PNAS* **71**: 2329–2331.

Konhauser, K. O., Phoenix, V. R., Bottrell, S. H., Adams, D. G., and Head, I. M., 2001, Microbial-silica interactions in Icelandic hot spring sinter: possible analogues for some Precambrian siliceous stromatolites, *Sedimentology* **48**: 415–433.

Lowe, D. R., 1994, Early environments: constraints and opportunities for early evolution. In: Bengston, S. (Ed.) *Early Life on Earth, Nobel symposium* **84**: 24–35.

Lowe, D. R., and Tice, M. M., 2004, Geologic evidence for Archean atmospheric and climatic evolution: fluctuating levels of $CO_2$, $CH_4$, and $O_2$ with an overriding tectonic control, *Geology* **32**: 493–496.

Luisi, P. L., 1998, About various definitions of life, *Origins of Life and Evolution of the Biosphere* **28**: 613–622.

McCall, G. J. H., 2003, A critique of the analogy between Archaean and Phanerozoic tectonics based on regional mapping of the Mesozoic-Cenozoic plate convergent zone in the Makran, Iran, *Precambrian Research* **127**: 5–17.

McCollom, T. M., and Seewald, J. S., 2006, Carbon isotope composition of organic compounds produced by abiotic synthesis under hydrothermal conditions, *Earth and Planetary Science Letters* **243**: 64–84.

Mileikowsky, C., Cucinotta, F., Wilson, J. W., Gladman, B., Horneck, G., Lindegren, L., Melosh, J., Rickman, H., Valtonen, M., and Zheng, J. Q., 2000, Natural transfer of viable microbes in space, part 1: from Mars to Earth and Earth to Mars, *Icarus* **145**: 391–427.

Moorbath, S., 2005, Oldest rocks, earliest life, heaviest impacts, and the Hadean–Archaean transition, *Applied Geochemistry* **20**: 819–824.

Oehler, D. Z., Robert, F., Mostefaoui, S., Meibom, A., Selo, M., and McKay, D. S., 2006, Chemical mapping of Proterozoic organic matter at submicron spatial resolution, *Astrobiology* **6**: 838–850.

Rasmussen, B., 2000, Filamentous microfossils in a 3235-million-year- old volcanogenic massive sulphide deposit, *Nature* **405**: 676–679.

Sagan, C., and Mullen, G., 1972, Earth and Mars: evolution of atmospheres and surface temperatures, *Science* **177**: 52–56.

Sherwood-Lollar, B., Westgate, T. D., Ward, J. A., Slater, G. F., and Lacrampe-Couloume, G., 2002, Abiogenic formation of alkanes in the Earth's crust as a minor source for global hydrocarbon reservoirs, *Nature* **416**: 522–524.

Siever, R., 1992, The silica cycle in the Precambrian, *Geochimica et Cosmochimica Acta* **56**: 3265–3272.

Tyler, S. T., and Barghoorn, E. S., 1963, Ambient pyrite grains in Precambrian cherts, *American Journal of Science* **261**: 424–432.

Uwins, P. J. R., Webb, R. I., and Taylor, A. P., 1998, Novel nano-organisms from Australian sandstones, *American Mineralogist* **83** (11–12 part 2): 1541–1550.

Van Kranendonk, M. J., Collins, W. J., Hickman, A., and Pawley, M. J., 2004, Critical tests of vertical vs. horizontal tectonic models for the Archaean East Pilbara Granite-Greenstone Terrane, Pilbara Craton, Western Australia, *Precambrian Research* **131**: 173–211.

Wacey, D., Kilburn, M. R., Stoakes, C. A., Aggleton, H., and Brasier, M. D., 2008a, Ambient inclusion trails: their recognition, age range and applicability to early life on earth. In: Dilek, Y., Furnes, H., and Muehlenbachs, K. (Eds.) *Links Between Geological Processes, Microbial Activities and Evolution of Life*, Springer, pp. 113–133.

Wacey, D., Kilburn, M. R., McLoughlin, N., Parnell, J., Stoakes, C. A., and Brasier, M. D., 2008b, Use of NanoSIMS to investigate early life on Earth: ambient inclusion trails in a *c.* 3400 Ma sandstone, *Journal of the Geological Society of London* **165**: 43–53.

Walde, P., 2005, *Prebiotic Chemistry: from simple amphiphiles to protocell models, topics in current chemistry* **259**, Springer, Berlin, 221 p.

Westall, F., and Folk, R. L., 2003, Exogenous carbonaceous microstructures in Early Archaean cherts and BIFs from the Isua Greenstone Belt: implications for the search for life in ancient rocks, *Precambrian Research* **126**: 313–330.

Westall, F., Boni, L., and Guerzoni, M. E., 1995, The experimental silicification of microbes, *Palaeontology* **38**: 495–528.

Williams, R.P. J., and Frausto da Silva, J. J. R., 1996, *The Natural Selection of the Chemical Elements*, Clarendon, Oxford, 646 p.

Xiao, S., and Knoll, A. H., 1999, Fossil preservation in the NeoProterozoic Doushanto phosphorite Lagerstatte, South China, *Lethaia* **32**: 219–240.

Zhang, Z., and Goloubic, S., 1987, Endolithic microfossils (cyanophyta) from early Proterozoic Stromatolites, Hebei, China, *Acta Micropalaeontoligica Sinica* **4**: 1–12.

# Chapter 2
# The Difficulties of Decoding Early Life

## 2.1 Introduction

We now know that the search for evidence of the earliest life of Earth focuses on finding morphological remains or body (micro)fossils, morphological trace fossils indicative of life, and/or chemical trace fossils indicative of biological metabolisms. This sounds relatively straightforward, but several difficulties plague these investigations. A sound understanding of these difficulties is necessary to prevent much frustration both in the field and in the laboratory.

## 2.2 Non-Biological Artefacts

The perennial difficulty with interpreting all 'microfossils' is that they comprise shapes (spheres, rods, filaments) that are difficult to distinguish from natural non-microbial mineral crystal habits that could grow under similar conditions. Trace fossils suffer from similar problems. Simple non-biological experiments demonstrate the ease with which microfossil-like artefacts can be generated by geologically relevant processes (Figs. A3, B92 and B93). It may need to be demonstrated, therefore, that candidate microfossil-like structures occupy their own discrete area of morphospace, which does not overlap with that occupied by relevant non-biological processes such as crystal growth. The problem here is that a basic understanding of both self-organizing structures (SOS) and complexity is essential if the early fossil record is to be decoded correctly (see pages 15–17 of introduction to this book by Brasier). Such an understanding shows us that complex structures do not require complex causes, as was appreciated nearly a century ago by the ingenious experiments of d'Arcy Thompson (1917). Complexity can arise naturally in any physico-chemical system that has the realms of 'chaotic' behaviour. Stromatolites do not escape the complexity argument either as Grotzinger and Rothman (1996) elegantly showed over a decade ago.

Direct inorganic synthesis of microfossil-like objects has been attained in the laboratory (e.g., Garcia-Ruiz et al., 2003). These objects are non-crystallographic, commonly curved or helical (see Fig. B92). Simple non-biological hydrocarbons

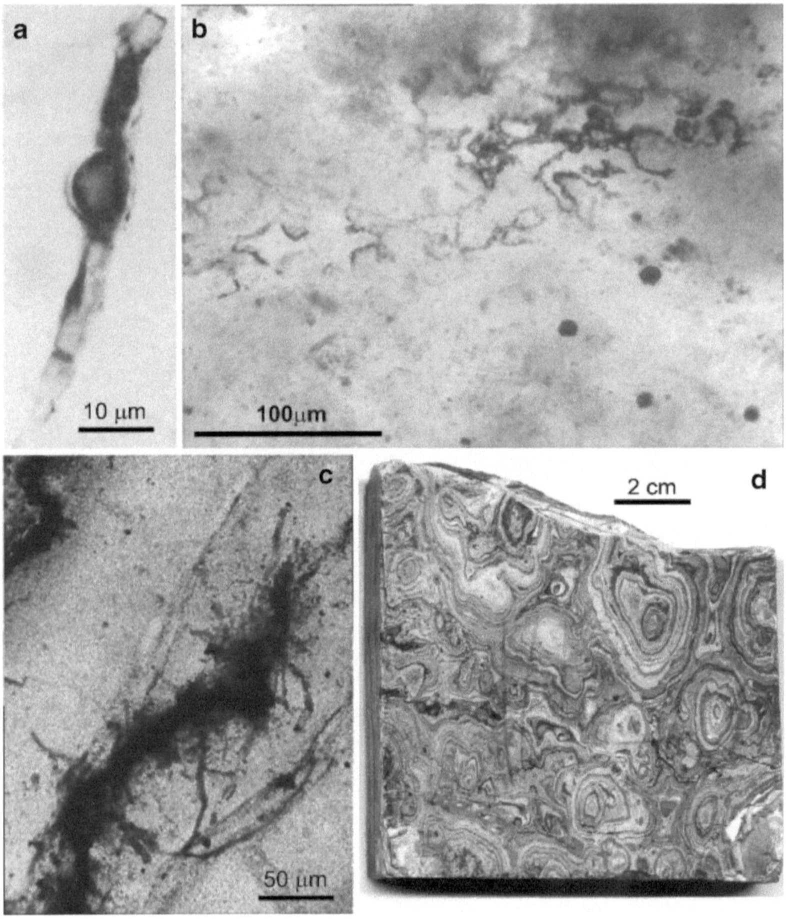


**Fig. A3** Examples of modern contamination and non-biological artefacts. (**a**) Microfossil found in a ~2,220 Ma dolomitic stromatolite from South Africa, originally interpreted as a ~2,220 Ma filamentous blue green algae (*Petraphera vivescenticula*; Nagy, 1974), but subsequently re-interpreted as a modern contaminant (Hofmann and Schopf, 1983). Image reproduced with permission from AAAS. (**b**) Thin section image of microfossil-like artefacts from the ~3,460 Ma Apex chert, Western Australia, caused by carbonaceous matter wrapping around crystal margins. (**c**) Thin section image of a modern contaminant in the ~3,460 Ma Apex chert. This takes the form of a rootlet that has penetrated micro-cracks in the chert. Such rootlets can penetrate distances of several metres. (**d**) Polished slab of a non-biological stromatolite-like structure. This is a plan view of the 'Taylor Stromatolite' (see p. 8) which contains hundreds of ~50 μm thick multicoloured, convex up laminae that form laterally linked domes and columns. It was formed as an overspray deposit that accumulated on a paint spray booth from an automobile production line (Images (**b–d**) are from the Oxford Palaeobiology Collections)

can condense onto these filaments and when these undergo subsequent diagenesis they can show remarkable morphological and chemical (e.g., kerogenous composition) similarity to putative Archean microfossils. As yet, these inorganic filaments have only been synthesised under alkaline, hydrothermal conditions, but this is not un-reconcilable with the conditions that may have been prevalent on the early Earth (see list on pages 30–31).

A simple example of mistaken identity comes from ~3,800 Ma meta-quartzite rocks from the Isua region of Greenland. In the late 1970s Pflug and colleagues described a suite of yeast like 'microfossils' (*Isuasphaera isua,* see page 128; Pflug and Jaeschke-Boye, 1979) from these rocks. They interpreted the simple structures as being biological based upon claims of cell walls, cell vacuoles, gas vacuoles, sheaths, budding structures and remnants of protoplasm. Re-examination of these rocks by Bridgewater et al. (1981) showed that the microfossil-like objects were in fact a mixture of limonite (hydrated iron oxide) stained fluid inclusions and stained quartz grain boundaries.

One of the great problems with claims for early life is the lack of independent corroboration. Too often a claim is made and accepted into the text book literature without any series of checks being carried out. This is particularly concerning given the fact that, in the few cases where checks have been carried out, corroboration has not been forthcoming. The most famous example of this is the description of a suite of 'microfossils' from the ~3,460 Ma Apex chert from the Pilbara of Western Australia (Schopf, 1993) that were taken as *the bona fide fossil evidence for life* for almost a decade before they were independently re-examined. When they were re-examined these 'microfossils' were re-interpreted as non-biological artefacts (Brasier et al., 2002, 2005) (Fig. A3b, and see pages 12–15 and 155–159 for details).

A comprehensive range of definitive non-biological artefacts are described in the final chapter of this book (pages 251–266) and users should familiarise themselves with such features before examining rocks under the optical or electron microscope.

## 2.3 Post-Depositional Contamination

Archean rocks have undergone long and complex histories, experiencing substantial modification from their original depositional state. The opportunity for contamination by later microbial material is vast. Rocks can be colonised by microbial communities long after their formation and vestiges of this colonization can be brought about both passively (e.g., transported and cemented into pore spaces and fractures by post-depositional fluids), and actively (e.g., endolithic bacteria boring into lithified sediments). Organic material can also be caught up in younger fluid inclusions, and in modern and ancient percolating ground water. This immediately renders any bulk rock analytical techniques less than ideal.

In some cases, post-depositional microbial matter can be recognised by the morphology of the mineral phase enclosing it. In siliceous rocks such as cherts, for example, the oldest generation of silica tends to be microcrystalline, formed from the early diagenetic transformation of primary aqueous opaline silica, and tends to form interlocking grains. This diagenetic transformation may be evidenced by the obliteration of the fine scale morphological features of some microfossils, whose features are smaller than the grain size of the microcrystalline quartz. A later generation of silica that rapidly fills primary cavities is known as chalcedony; this follows the contours of the cavity to form botryoidal masses. Silica which infills even later veins and fractures tends to form angular grains which are much larger than the early microcrystalline variety. A microfossil found in chalcedonic or vein fill quartz is still a microfossil (and may indeed still be quite ancient), but it is not syngenetic (i.e., the same age) with the depositional age of the rock. A microfossil found in microcrystalline quartz is more likely to be syngenetic with its host sediment, or at worst introduced very soon after deposition, during earliest diagenesis.

Examples of modern contamination of Archean rocks are numerous (see pages 251–266 for comprehensive examples). A couple of illustrative examples will demonstrate the point here. Nagy (1974) reported blue-green algae and other filamentous and coccoid microfossils from the Proterozoic Malmani Dolomite in South Africa (Fig. A3a). A re-examination of these thin sections (Hofmann and Schopf, 1983), however, showed the microfossils to be red-brown in colour and occurring in cylindrical void spaces within the original rock fabric. These are now best interpreted as modern endolithic microbes that have bored into the carbonate matrix. In the Isua region of Greenland, an area synonymous with the search for Earth's oldest life, Westall and Folk (2003) discovered numerous fossilized endolithic microbes and carbonaceous matter in ~3,700 Ma banded iron formation (BIF) samples. The carbonaceous matter is located on grain surfaces in the proximity of cracks, indicating it has been probably been washed in by rainwater very recently (the last 8,000 years, because before this time the area was completely ice bound). Filamentous and coccoid microorganisms, together with EPS were found on fracture surfaces of rock chips and along grain boundaries viewed in a scanning electron microscope, again indicating a modern endolithic origin (also see Fig. B94). Although some carbonaceous matter, in the form of graphite, is syngenetic with deposition of these rocks, the discovery of modern endoliths casts serious doubt upon previous bulk rock carbon isotope measurements from Greenland. These $\delta^{13}C$ values may be a mixture of this modern endolithic signature and the original ~3,700 Ma graphite signal.

## 2.4 The Pros and Cons of the 'Principle of Uniformity'

Observations made upon modern and observable natural phenomena can also be applied to the ancient rock record. This approach, known as the 'Principle of Uniformity', has been a guiding principle within the Earth Sciences since the time of James Hutton (1785). A classic example of this is the 'Early Eden Hypothesis' as applied to the early Earth. The main claims made by this hypothesis have now dominated thinking for several decades of space exploration, international debate, media coverage and teaching about the early history of life. This Early Eden Hypothesis takes familiar and habitable environments in which primitive microbes abound today, such as Bahamian tropical lagoons, Mono Lake in California, or

Shark Bay in Western Australia and then uses these to make predictions about the surface of the early Earth. This is, of course, a tried and tested method and can be argued to work reasonably well when applied to the rock record from the Quaternary back into the early Proterozoic or even the late Archean (2,800–2,500 Ma). However, the Principle of Uniformity may be pushed beyond its limits when extended back into the early Archean. In its most extreme expression, the Early Eden Hypothesis predicts the presence, on the early Earth, of continents, subduction zones, carbonate platforms, an oxygenated atmosphere and oxygenic photosynthesis. Examination of the earliest sedimentary rocks, however, coupled with an ever increasing understanding about the nature of the solar system, suggests that Lyell's much vaunted Principle of Uniformity may be misleading us somewhat (Rose et al., 2006). It is better to remember the warnings of Sir Francis Bacon here: "*The subtlety of nature is greater many times over than the subtlety of the senses and understanding; so that all those specious meditations, speculations, and glosses in which men indulge are quite from the purpose, only there is no one by to observe it*" (Bacon, 1620). In other words, we need to remain aware of the huge gaps in our understanding at this time, and the huge potential differences in the atmosphere, hydrosphere and lithosphere at this time (refer to list on pages 30–31).

Given the radically different boundary conditions acting upon the early Earth, it appears that the Earth's endogenic energy was potentially a much greater source of energy for the early biosphere than the solar energy of our star, the Sun. A first consequence is that the highly metaliferous crust of the early Earth, when combined with enormous outflows of energy emanating from hydrothermal and volcanic systems, is likely to have played a significant role in both the genesis and sustenance of the earliest forms of life. This message is also delivered to us by the discovery of thriving life forms around black smokers and modern deep sea vents, together with theoretical and chemical studies confirming that a "hydrothermal cradle for life" is perfectly plausible (e.g., Shock, 1990; Stetter, 1996). A second consequence of this view of the early Earth as a distant planet is that oxygenic photosynthesis need not have been the keystone to all other forms of life, as it might seem to us today.

The 'Principle of Uniformity' suffers from two further fundamental weaknesses in the early Archean: preservational bias; and reconciling stratigraphic and sedimentologic time scales. It is well known that some environments are highly likely to be preserved and these may come to dominate the sedimentary rock record. Others lack known mechanisms for preservation and therefore remain either exceptional or unrecognized within the rock record. Representing the former are coastal marshes and dune fields, tidal flats, lowland lakes, swamps, and floodplains at or near sea level. Representing the latter are mountain environments and high-energy coasts where erosion removes the rocks; or deep ocean abyssal plains and trenches, which are rarely preserved because they are severely deformed by tectonic processes which may very rarely emplace them on to the continents. We must therefore accept that our reconstructions of ancient environments are limited to a much smaller range of physical settings than those that actually existed in the past.

The second weakness relates to the problem of reconciling stratigraphic and sedimentologic time scales. For example, how can short-term processes of sediment deposition in active environments that operate at time scales of days to months to tens of years, be reconciled in the rock record with geochronologies (measured by radioactive decay, see Section 5.3) measured in thousands of years to tens of millions of years? Our tendency is to time-average the rocks between chronologic tie-points. This greatly understates the staccato aspect of the rock record, which may more accurately be seen as a stack of ordered snapshots rather than as a movie reel. Gaps predominate, but it is not straightforward to see how these gaps are distributed within a given geological outcrop.

Despite these weaknesses, comparisons to modern day phenomena certainly deserve a place in studies of early life. Modern analogues help to constrain and sometimes quantify the range of habitats and metabolisms microbes can utilise, and the types of isotopic fractionations and chemical enrichments that may be indicative of life. For example work on modern hydrothermal deep sea and hot spring environments has extended the known temperatures and pressures that microbes can metabolise under and has allowed geologists working on early life to investigate rocks that may otherwise have gone unchecked.

## 2.5 A Benchmark for Microfossils and Stromatolites

In the course of investigating the Archean fossil record many researchers have sought a benchmark against which to assess claims for the earliest signs of life. The Proterozoic Gunflint Chert microfossil assemblage has been advanced as such a benchmark (e.g., Moorbath, 2005; Rose et al., 2006); the first known Precambrian microfossils were discovered in the 1960s from stromatolitic cherts of the Gunflint Formation in the Lake Superior region of Canada (Barghoorn and Tyler, 1965). These microfossils are some ~1,878 Ma in age (Fralick et al., 2002) and rightfully created much excitement on their discovery. They predate the Cambrian explosion and rapid radiation of metazoan life by ~1,300 Ma (Knoll, 2003) and were discovered at a time when the Precambrian time interval was thought to be devoid of fossils.

The Gunflint assemblage contains vast numbers of individual microfossils that have been placed into more than 20 different genera. As seen in Fig. A4a–e, the organisms are of diverse and rather complex morphology. The abundance, diversity and complexity of these microfossils has resulted in widespread acceptance of the Gunflint microfossils as *bona fide* examples of life almost 2 billion years ago. The Gunflint biota is not alone as an example of remarkable preservation in the Proterozoic. Other good examples from the Proterozoic that are useful benchmarks for any early Archean finds include:

- ~2,400 Ma Kuruman Iron Formation, Transvaal Supergroup, South Africa (Klein et al., 1987) where filamentous microfossils (interpreted as cyanobacterial sheaths) are preserved in both chert and dolomite.
- ~2,000 Ma Frere Formation, Nabberu Basin, Western Australia (Walter et al., 1976) where microfossils are found preserved in a stromatolitic iron formation.

- ~2,000 Ma Duck Creek Dolomite of Western Australia (Knoll and Barghoorn, 1976; Knoll et al., 1988) where microfossils in different stages of degradation prior to fossilization can be seen. These are preserved in black chert lenses and many are morphologically similar to the Gunflint biota.
- ~2,000 Ma Dahongyu Formation, Hebei Province, China (Yun, 1984) where Gunflint-like microfossils are found in silicified stratiform stromatolites.
- ~2,000 Ma Tyler Formation of northern Michigan (Cloud and Morrison, 1980).
- ~1,900 Ma Belcher Supergroup, Canada (Hofmann, 1976) where microfossils occur in a silicified stromatolitic dolomite unit. Eighteen genera of filamentous and coccoidal microfossils are described and illustrated in this report (e.g., Fig. A4f–h).
- ~1,500 Ma Amelia Dolomite (Muir, 1976), Balbirini Dolomite (Oehler, 1979) and Barney Creek Formation (Oehler, 1977) of the McArthur Group, Australia.
- ~1,000 Ma Shorikha and Burovaya Formations, northeastern Siberia (Sergeev, 2001) where both prokaryotic and eukaryotic microfossils are preserved in chert lenses.
- ~850 Ma Bitter Springs Formation, Central Australia (Schopf, 1968; Schopf and Blacic, 1971) where the microfossils are preserved in bedded carbonaceous chert. Numerous filamentous (e.g., Fig. A4i) and coccoidal prokaryotes (and some eukaryotes) are described and illustrated in these reports.

This is by no means an exhaustive list (you will find many more examples simply by flicking though the references in the reports outlined above) but it does provide a good range of representative morphologies and states of preservation with which to compare any putative early Archean microfossils.

The Gunflint Formation also serves as a good benchmark for stromatolites since the microfossils previously described occur in stromatolitic ironstones (Awramik and Semikhatov, 1978). However, the benchmark for biological stromatolites can safely be pushed back some ~1,000 Ma, using two undisputed fossil stromatolite reefs from the ~2,900 Ma Steep Rock Group in north-west Ontario and the ~2,700 Ma Tumbiana Formation from Western Australia.

In the Steep Rock Group an entire carbonate platform is preserved with branched columnar stromatolites and very large, spectacular domal stromatolites up to about 3 m in diameter (Fig. A5b). The shear size and lateral extent of these structures, combined with a well understood geological context, and numerous biological features (see pages 51–52) on both the macro- and micro-scale means that a non-biological formation mechanism can be rejected. Carbon isotopes have also been measured on kerogenous laminae at sub-millimetre scale spatial resolution. These values (–21.6‰ to –30.6‰ PDB) lend supporting evidence to the biological nature of these stromatolites (Grassineau et al., 2006).

The Meentheena carbonate member of the Tumbiana Formation contains widespread columnar and bulbous domical stromatolites (Fig. A5a, f). Some of the larger columns and bulbs are several tens of centimetres in diameter. Importantly, the Tumbiana stromatolites display both the same macrostructure and microstructure as modern varieties from Shark Bay (Fig. A5c, d); the same cannot be said for >3,000 Ma putative stromatolite examples. A recent study of the Tumbiana stromatolites has added geochemical arguments for their biogenicity to the more obvious morphological ones

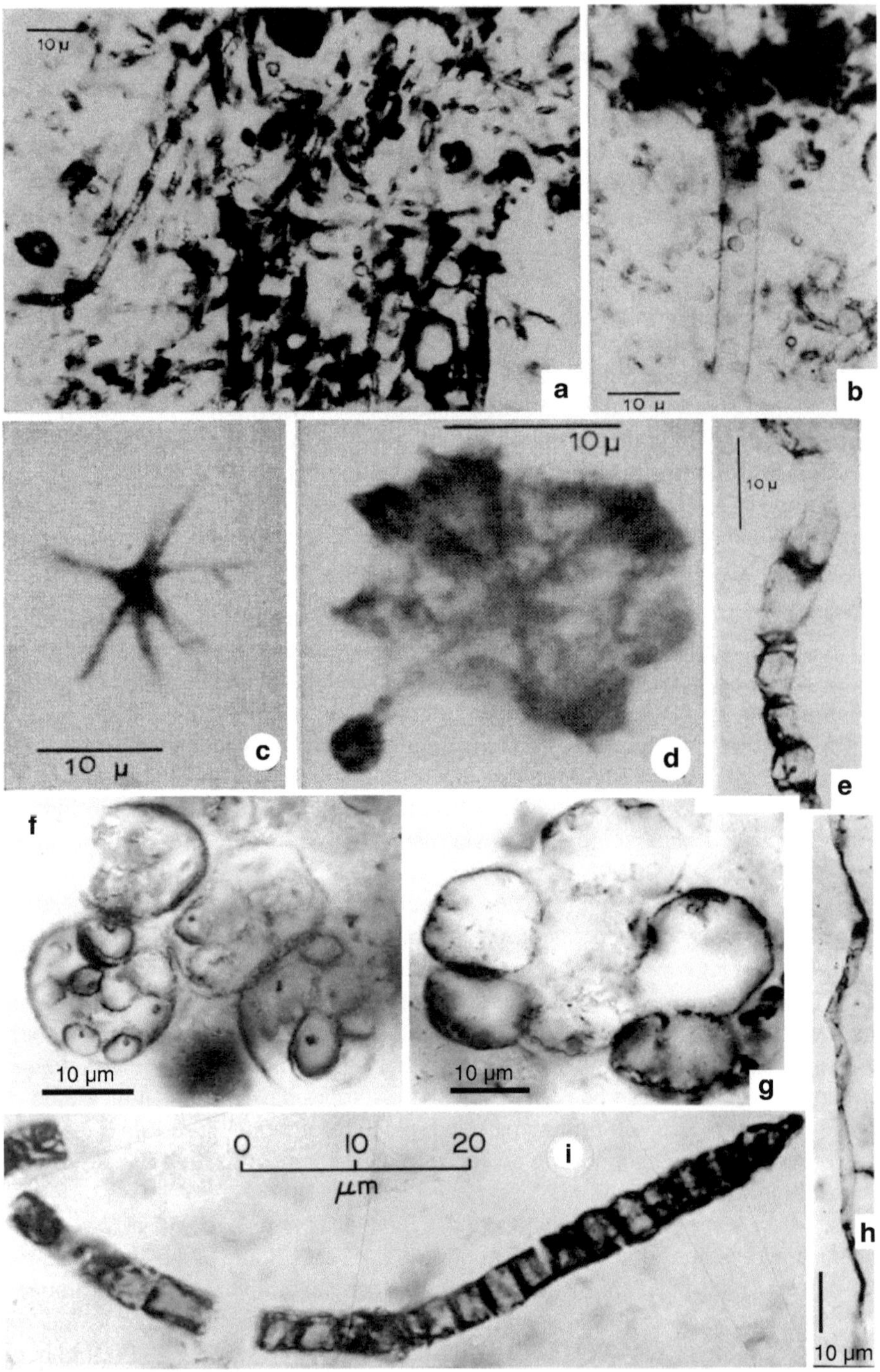


**Fig. A4** Examples of well preserved and widely accepted Proterozoic microfossils that can be used as a benchmark for putative early Archean examples. (**a–e**) Diverse microfossils from the ~1,900 Ma Gunflint Formation, Ontario (Barghoorn and Tyler, 1965, reproduced with permission

**Fig. A5** Stromatolites old and new. (**a**) Well preserved columnar stromatolites of undisputed biological origin from the ~2,700 Ma Meentheena carbonate member of the Tumbiana Formation, Western Australia. Pen for scale. (**b**) Giant domal stromatolites of undisputed biological origin from the Steep Rock Group, Ontario. Person for scale. (**c**) Modern stromatolites from Shark Bay, Western Australia. (**d**) Cut slab through a Shark Bay stromatolite – note the very similar columnar morphology to the Tumbiana example. Lens cap for scale. (**e**) Thin section image of the microstructure of a modern stromatolite from Lake Thetis, Western Australia. (**f**) Thin section image of the microstructure of a Tumbiana stromatolite. (**g**) Thin section image of a putative biological stromatolite from the ~3,400 Ma Strelley Pool Formation, Western Australia. Note the similarity between the microstructures observed in the modern and Tumbiana examples, and lack of similarity of the Strelley Pool example ((**a**) Courtesy of the Geological Survey of Western Australia; (**b–d**) courtesy of Martin Brasier; (**e–g**) taken by the author)

**Fig. A4** (continued) from AAAS). (**f–h**) Coccoidal and filamentous microfossils from the ~1,900 Ma Belcher Supergroup, Canada (Hofmann, 1976, reproduced with permission from SEPM). (**i**) Filamentous microfossil from the ~850 Ma Bitter Springs Formation, Central Australia (Schopf and Blacic, 1971, reproduced with permission from SEPM)

(Lepot et al., 2008). This study reports clusters of hundreds of microbe-shaped organic globules associated with tiny spheroids of aragonite, a relationship that is replicated in modern microbial sediments (Dupraz et al., 2004). High resolution geochemistry confirms their antiquity and the presence of a number of carbonaceous functional groups found in modern bacteria. These kind of studies are now required, in addition to morphological comparisons, to demonstrate the biogenicity of any >3,000 Ma stromatolites.

## Recommended Reading

Awramik, S. M., and Semikhatov, M. A., 1978, The relationship between morphology, microstructure and microbiota in three vertically intergrading stromatolites from the Gunflint Iron Formation, *Canadian Journal of Earth Sciences* **16**: 484–495.

Bacon, F., 1620, *Novo Organum.*

Barghoorn, E. S., and Tyler, S. A., 1965, Microorganisms from the Gunflint Chert, *Science* **147**: 563–577.

Brasier, M. D., Green, O. R., Jephcoat, A. P., Kleppe, A. K., Van Kranendonk, M. J., Lindsay, J. F., Steele, A., and Grassineau, N. V., 2002, Questioning the evidence for Earth's oldest fossils, *Nature* **416**: 76–81.

Brasier, M. D., Green, O. R., Lindsay, J. F., McLoughlin, N., Steele, A., and Stoakes, C., 2005, Critical testing of Earth's oldest putative fossil assemblage from the ~3.5 Ga Apex Chert, Chinaman Creek, Western Australia, *Precambrian Research* **140**: 55–102.

Bridgewater, D., Allaart, J. H., Schopf, J. W., Klein, C., Walter, M. R., Barghoorn, E. S., Strother, P., Knoll, A. H., and Gorman, B. E., 1981, Microfossil-like objects from the Archaean of Greenland: a cautionary note, *Nature* **289**: 51–53.

Cloud, P., and Morrison, K., 1980, New microbial fossils from 2 Gyr rocks in northern Michigan, *Geomicrobiology Journal* **2**: 161–178.

Dupraz, C., Visscher, P. T., Baumgartner, L. K., and Reid, R. P., 2004, Microbe-mineral interactions: early carbonate precipitation in a hypersaline lake (Eleuthera Island, Bahamas), *Sedimentology* **51**: 745–765.

Fralick, P., Davis, D. W., and Kissin, S. A., 2002, The age of the Gunflint Formation, Ontario, Canada: single zircon U-Pb age determinations from reworked volcanic ash, *Canadian Journal of Earth Sciences* **39**: 1085–1091.

Garcia-Ruiz, J. M., Hyde, S. T., Carnerup, A. M., Christy, A. G., Van Kranendonk, M. J., and Welham, N. J., 2003, Self-assembled silicacarbonate structures and detection of ancient microfossils, *Science* **302**: 1194–1197.

Grassineau, N. V., Abell, P., Appel, P. W. U., Lowry, D., and Nisbet, E. G., 2006, Early life signatures in sulfur and carbon isotopes from Isua, Barberton, Wabigoon (Steep Rock), and Belingwe Greenstone Belts (3.8 to 2.7 Ga). In: Kesler, S. E., and Ohmoto, H. (Eds.) *Evolution of Early Earth's Atmosphere, Hydrosphere and Biosphere – Constraints from Ore Deposits, Geological Society of America Memoir* **198**: 33–52.

Grotzinger, J. P., and Rothman, D. H., 1996, An abiotic model for stomatolite morphogenesis, *Nature* **383**: 423–425.

Hofmann, H. J., 1976, Precambrian microflora, Belcher Islands, Canada: significance and systematics, *Journal of Paleontology* **50**: 1040–1073.

Hofmann, H. J., and Schopf, J. W., 1983, Early Proterozoic microfossils. In: Schopf, J. W. (Ed.) *Earth's Earliest Biosphere, Its Origin and Evolution*, Princeton University Press, Princeton, NJ, pp 321–360.

Hutton, J., 1785, *Theory of the Earth*. An investigation of the laws observable in the composition, dissolution, and restoration of land upon the globe. *Transactions of the Royal Society of Edinburgh*, vol. I, Part II, pp. 209–304.

Klein, C., Beukes, N. J., and Schopf, J. W., 1987, Filamentous microfossils in the early Proterozoic Transvaal Supergroup: their morphology, significance, and paleoenvironmental setting, *Precambrian Research* **36**: 81–94.

Knoll, A. H., 2003, *Life on a Young Planet: The First Three Billion Years of Evolution on Earth*, Princeton University Press, Princeton, NJ, 277 p.

Knoll, A. H., and Barghoorn, E. S., 1976, A Gunflint-type microbiota from the Duck Creek dolomite, Western Australia, *Origins of Life* **7**: 417–423.

Knoll, A. H., Strother, P. K., and Rossi, S., 1988, Distribution and diagenesis of microfossils from the lower Proterozoic Duck Creek Dolomite, Western Australia, *Precambrian Research* **38**: 257–279.

Lepot, K., Benzerara, K., Brown, G. E., and Philippot, P., 2008, Microbially influenced formation of 2,724-million-year-old stromatolites, *Nature Geoscience* **1**: 118–121.

Moorbath, S., 2005, Oldest rocks, earliest life, heaviest impacts, and the Hadean–Archaean transition, *Applied Geochemistry* **20**: 819–824.

Muir, M. D., 1976, Proterozoic microfossils from the Amelia Dolomite, McArthur Basin, Northern Territory, *Alcheringa* **1**: 143–159.

Nagy, L. A., 1974, Transvaal stromatolites: first evidence for the diversification of cells about $2.2 \times 10^9$ years ago, *Science* **183**: 514–516.

Oehler, J. H., 1977, Microflora of the HYC pyritic shale member of the Barney Creek Formation (McArthur Group), middle Proterozoic of northern Australia, *Alcheringa* **1**: 315–349.

Oehler, D. Z., 1979, Microflora of the middle Proterozoic Balbirini Dolomite (McArthur Group) of Australia, *Alcheringa* **2**: 269–310.

Pflug, H. D., and Jaeschke-Boyer, H., 1979, Combined structural and chemical analysis of 3,800-Myr-old microfossils, *Nature* **280**: 483–486.

Rose, E. C., Mcloughlin, N., and Brasier, M. D., 2006, Ground truth: the epistemology of searching for the earliest life on Earth. In: Seckbach, J. (Ed.), *Life as We Know It: Cellular Origin, Life in Extreme Habitats and Astrobiology* **10**, Springer, Dordrecht, The Netherlands, 650 p.

Schopf, J. W., 1968, Microflora of the Bitter Springs Formation, Late Precambrian, Central Australia, *Journal of Palaeontology* **42**: 651–688.

Schopf, J. W., 1993, Microfossils of the Early Archaean Apex Chert: new evidence for the antiquity of life, *Science* **260**: 640–646.

Schopf, J. W., and Blacic, J. M., 1971, New microorganisms from the Bitter Springs Formation (Late Precambrian) of the North-Central Amadeus Basin, Australia, *Journal of Paleontology* **45**: 925–960.

Sergeev, V. N., 2001, Paleobiology of the Neoproterozoic (Upper Riphean) Shorikha and Burovaya silicified microbiotas, Turukhansk Uplift, Siberia, *Journal of Palaeontology* **75**: 427–448.

Shock, E. L., 1990, Geochemical constraints on the origin of organic compounds in hydrothermal systems, *Origins of Life and Evolution of the Biosphere* **20**: 331–367.

Stetter, K. O., 1996, Hyperthermophiles in the history of life. In: Bock, G. R., and Goode, J. A. (Eds.) *Evolution of Hydrothermal Ecosystems on Earth (and Mars?)*, Wiley, Chichester, pp. 1–18.

Thompson, D'A. W., 1917, *On Growth and Form*, Cambridge University Press, Cambridge.

Walter, M. R., Goode, A. D. T., and Hall, W. D. M., 1976, Microfossils from a newly discovered Precambrian stromatolitic iron formation in Western Australia, *Nature* **261**: 221–223.

Westall, F., and Folk, R. L., 2003, Exogenous carbonaceous microstructures in Early Archaean cherts and BIFs from the Isua Greenstone belt: implications for the search for life in ancient rocks, *Precambrian Research* **126**: 313–330.

Yun, Z., 1984, A Gunflint type of microfossil assemblage from early Proterozoic stromatolitic cherts in China, *Nature* **309**: 547–549.

# Chapter 3
# Establishing the Criteria for Early Life on Earth

## 3.1 Introduction

A significant but widely ignored problem in early life studies concerns our reliance upon inductive lines of reasoning. More specifically, there has tended to be too much reliance upon evidence that is 'consistent with' microbes, without falsifying or rejecting (*sensu* Popper, 1959) other possible non-biological scenarios that may likewise be consistent. We have tended to ask 'what do these structures remind us of', rather than 'what are these structures'? Recognition of the need for testing a null hypothesis of a non-biological origin for the earliest fossil evidence therefore forces us to face up to, and overcome, this very human tendency. Arguments of the kind: 'absence of evidence is not evidence of absence' are also highly insecure in the high-stakes search for early life, and are incompatible with evolutionary studies of all kinds.

Attempts have been made to establish sets of more rigid criteria that any claims for early life should adhere to. These are here divided into criteria used to demonstrate the antiquity of a given structure and those used to demonstrate the biogenicity of a structure.

## 3.2 Antiquity Criteria

An exploration of the criteria needed to establish the antiquity of putative microfossils or trace fossils is the first step when studying early life, given that an investigation of their biogenicity is contingent upon a full understanding of their age. Proving that a microfossil or trace fossil is of undoubted Archean age is rather more difficult than proving the age of a younger fossil. An Archean rock has undergone a long and complex geological history, possibly with many stages of burial, heating and exhumation, meaning that the opportunities for post-depositional contamination are plentiful. The criteria used to attempt to untangle these processes are given below (summarised from Brasier et al., 2004, 2005; Buick, 1990; Hofmann, 2004; McLoughlin et al., 2007; Schopf and Walter, 1983; Sugitani et al., 2007; Westall and Folk, 2003)

D. Wacey, *Early Life on Earth: A Practical Guide*,

### 3.2.1 General Antiquity Criteria

(a) Structures must occur in rocks of known provenance; i.e., detailed location information must be presented so that independent re-sampling is possible.
(b) Structures must occur in rocks of demonstrable or established (Archean) age; i.e., the host rock must be dated directly by radiometric techniques, or the age of the rocks can be accurately inferred by correlation to nearby rocks that have been dated.
(c) Structures must be indigenous to the primary fabric of the host rock; i.e., they must be physically embedded within the rock, not products of sample collection or preparation. They should, therefore, be present in petrographic thin sections of the rock. Other identification techniques such as acid maceration and acid etching are valuable accessory techniques but may accidentally incorporate post depositional contaminants.
(d) Structures must be syngenetic with the primary fabric of the host rock; i.e., they must not have been introduced by ancient or modern post-depositional fluids.
(e) Following from (d); any structures found within meta-stable mineral phases, void filling cements, veins, or cross-cutting fabrics must be viewed with extreme caution.
(f) Structures should not occur in high grade metamorphic rocks, because delicate organic structures will not survive these extremes of pressure and/or temperature; the likelihood of non-biological artefacts in such rocks is substantially increased.
(g) The geological context of the host rock must be fully understood at a range of scales; i.e., the host unit must show geographical extent and fit logically within the regional geological history.

### 3.2.2 Additional Antiquity Criteria Specific to Microfossils

(a) Potential microfossils should not be significantly different in colour from that of particulate carbonaceous material in the remainder of the rock matrix. For example, brown 'microfossils' in a largely black carbonaceous chert would immediately be suspicious.
(b) There should be evidence for organo-sedimentary interaction, e.g., sediment grains trapped or supported by fossils, coatings of distinctive composition or texture precipitated around the fossils, or perhaps alternating layers of prostrate and erect filaments in stromatolite-like sediments.

### 3.2.3 Additional Antiquity Criteria Specific to Trace Fossils

(a) Trace fossils should be concentrated in detrital grains or primary rock matrix, not around later conduits for fluid or microbial entry.

(b) Trace fossils should be cross cut by later stage veins and fractures (if present), e.g., fractures that pass through many grains within the rock.
(c) Trace fossils should be filled with a mineral phase that is capable of surviving the history of burial and heating that the rock unit is known to have endured.

## 3.3 Biogenicity Criteria

The problem of demonstrating that a certain microstructure in a rock more than 3 billion years old is biological is not a trivial one. The most robust claims for early life are based upon mutually supporting lines of evidence drawn from the disciplines of geological mapping, geochemistry, and palaeobiology. Compliance with the following criteria (summarised from Brasier et al., 2004, 2005; Buick, 1990; Hofmann, 2004; McLoughlin et al., 2007; Schopf and Walter, 1983; Sugitani et al., 2007) should minimize the chance of interpreting a non-biological structure as biological.

### *3.3.1 General Biogenicity Criteria*

(a) Structures should exhibit biological morphology that can be related to extant cells, sheaths, traces of activity or waste products. Ideally life cycle variants should be identifiable (reproductive stages), comparable to that found in morphologically similar modern or fossil microorganisms.
(b) More than a single step of biology-like processing should be evident. These steps may take the form of biominerals (e.g., pyrite), geochemical fractionations of isotopes (e.g., carbon and sulphur), specific organic compounds (e.g., hopanoid biomarkers) or distinctive elemental ratios.
(c) Structures should occur within a geological context that is plausible for life; i.e., at temperatures and pressures that extant organisms are known to survive.
(d) Structures should fit within a plausible evolutionary context.
(e) Structures should be abundant and ideally occur in a multi-component assemblage.
(f) Following from (e), ideally they should show colonial/community behaviour.
(g) Following from (f), a preferred orientation indicating a role in the formation of biofabrics would be an additional bonus criterion.

### *3.3.2 Additional Biogenicity Criteria Specific to Microfossils*

(a) Microfossils should ideally be composed of kerogenous carbon. However, if mineralised this should be a result of microbially mediated precipitation. Later mineral replacement of carbonaceous material may also be permissible but then doubts upon antiquity will be raised.

(b) Microfossils should be largely hollow. Cell walls and sheaths are by far the most likely parts of the microbe to be preserved; cellular constituents are rarely preserved in more modern examples. Mineral artefacts are unlikely to be hollow.
(c) Ideally the microfossils should show some sort of cellular elaboration; e.g., not just smooth cell walls.
(d) Microfossils should show taphonomic degradation; i.e., collapse of cells, folding of films, fracturing. This may not occur in exceptional preservational circumstances, for example, in situ rapid silicification of living communities.
(e) The object must exceed the minimum size for independently viable cells (~0.25 μm diameter). Note: The recent discovery of nano-bacteria may modify this criterion to even smaller sizes.
(f) Microfossils should be demonstrably dissimilar from potentially co-existing non-biological organic bodies (e.g., self organising spherulitic structures), and should occupy a restricted biological morphospace.
(g) Evidence of extra-cellular polymeric substances surrounding the putative microfossils would be an added bonus criterion.

### *3.3.3 Additional Biogenicity Criteria Specific to Trace Fossils*

(a) Structures should show preferential exploitation of certain substrates or horizons. For example, those that are rich in trace metals utilized by microbial metabolisms, or those that contain structural defects and weaknesses that facilitate microboring.
(b) Structures should show enrichments of biologically important elements. For example, carbon and/or nitrogen enriched linings, or bio-mineral infillings.
(c) Endolithic microborings should show preferred growth orientations; i.e., they should penetrate from the outside of a grain and grow inwards, and may also cluster on one side of the grain.
(d) Structures should be demonstrably dissimilar from co-existing non-biological etch pits and cracks; i.e., they should be circular to elliptical in cross section, and be of a restricted range of diameters.
(e) If the structures are branched, ideally there should not be a change of diameter of the structure at the branching point.
(f) The area immediately surrounding a trace fossil should ideally show depletion in biologically important elements.

## 3.4 The Problem of Stromatolites

Stromatolites are a particularly troublesome group of structures for the study of early life. Even the definition of a stromatolite is controversial. 'Stromatolite' may be simply a descriptive term without any indication of biological involvement:

> Attached, laminated, lithified sedimentary growth structures, accretionary away from a point or limited surface of initiation (Semikhatov et al., 1979)

More commonly, however, the involvement of biology is assumed when a structure is termed a 'stromatolite':

> An organo-sedimentary structure produced by sediment trapping, binding, and/or precipitation as a result of the growth and metabolic activity of microorganisms, principally cyanobacteria (Awramik et al., 1976)

In the early Archean, ***the involvement of biology in stromatolite formation cannot be assumed***; each structure must be critically investigated, just like putative microfossils and trace fossils, working from a set of criteria. Consequently the assessment of stromatolite biogenicity in the Archean is notoriously difficult and enigmatic. A critique of the current criteria is summarised below (modified from McLoughlin, 2006; original criteria from Buick et al., 1981; Hofmann, 2000).

(a) The structures must occur in undoubted sedimentary or meta-sedimentary rocks. A viable sedimentary environment is a necessary first condition to demonstrate the biogenicity of a stromatolite.
(b) It must be demonstrated that the structures are syn-sedimentary. It is necessary to exclude soft sediment deformation (e.g., Lowe, 1994) and/or later structural deformation as contributing to the resulting stromatolite morphology.
(c) There should be a preponderance of convex upwards structures. This is a very qualitative criterion and is neither necessary nor sufficient to demonstrate stromatolite biogenicity. Non-biological self organising structures such as agate crusts can exhibit convex-upwards morphologies. It is also misleading to imply a phototropic component to stromatolite growth, as purely chemotropic structures may have been more abundant in the early biosphere.
(d) Laminae should thicken over the crests of flexures. This qualitative criterion is designed to exclude non-biological, chemical crusts that are widely believed to exhibit laterally uniform thickness' (known as isopachous laminae, see page 10, and Pope and Grotzinger, 2000).
(e) If the structures are laminated the laminations should be wavy, wrinkled and/or have several orders of curvature. Again this qualitative criterion is designed to exclude more uniform precipitated crusts, but no limits are placed on the extent of 'crinkliness' or 'curvature' which are controlled by sedimentary rheology and the degree of diagenetic modification. Also non-biological structures can display very ramified morphologies (e.g., Grotzinger and Rothman, 1996).
(f) Microfossils or trace fossils should be present within the structures. This is far too rigid a criterion as the preservation potential of microbial remains is extremely low, such that this criterion would exclude more than 90% of described fossil stromatolites and furthermore, the majority of recent stromatolites only contain microorganisms in their outermost layers (Grey et al., 1999). Not withstanding this, the presence of microfossils in a stromatolite does not confirm their active role in formation of the structure, as they may simply have been passively entombed by the accreting structure.

(g) Changes in composition of the microfossil assemblages should be accompanied by morphological changes in the stromatolite. This is an extension of criterion (f) and is extremely prescriptive as only a few instances are known where this criterion is satisfied, for example, Awramik (1976) from the Gunflint chert and Seong-Joo and Golubic (1999) from the middle Proterozoic of China.
(h) The fossils or trace fossils must be organised in a manner indicating trapping, binding or precipitation of sediment by the living microorganisms. Again this would be desirable but is somewhat over-optimistic. Tufted microbial filaments, fenestrae and micropores created by the growth and decay of now absent microbes would be useful but are only found when diagenesis is minimal (e.g., Turner et al., 2000).

## Recommended Reading

Awramik, S. M., 1976, The relationship between morphology, microstructure and microbiota in three vertically intergrading stromatolites from the Gunflint Iron Formation, *Canadian Journal of Earth Sciences* **16**: 484–449.

Awramik, S. M., Margulis, L., and Barghoorn, E. S., 1976, Evolutionary processes in the formation of stromatolites. In: Walter, M. R. (Ed.) *Stromatolites*, Elsevier, Amsterdam, pp. 149–162.

Brasier, M. D., Green, O. R., and Mcloughlin, N., 2004, Characterization and critical testing of potential microfossils from the early Earth: the Apex 'microfossil debate' and its lessons for Mars sample return, *International Journal of Astrobiology* **3**: 1–12.

Brasier, M. D., Green, O. R., Lindsay, J. F., McLoughlin, N., Steele, A., and Stoakes, C., 2005, Critical testing of Earth's oldest putative fossil assemblage from the ~3.5 Ga Apex Chert, Chinaman Creek, Western Australia, *Precambrian Research* **140**: 55–102.

Brasier, M. D., McLoughlin, N., and Wacey, D., 2006, A fresh look at the fossil evidence for early Archaean cellular life, *Philosophical Transactions of the Royal Society B* **361**: 887–902.

Buick, R., 1984, Carbonaceous filaments from North Pole, Western Australia: are they fossil bacteria in Archaean stromatolites? *Precambrian Research* **24**: 157–172.

Buick, R., 1988, Carbonaceous filaments from North Pole, Western Australia: are they fossil bacteria in Archaean stromatolites? A reply, *Precambrian Research* **39**: 311–317.

Buick, R., 1990, Microfossil recognition in Archaean rocks: an appraisal of spheroids and filaments from 3500 M.Y old chert-barite at North Pole, Western Australia, *Palaios* **5**: 441–459.

Buick, R., Dunlop, J. S. R., and Groves, D. I., 1981, Stromatolite recognition in ancient rocks: an appraisal of irregularly laminated structures in an early Archaean chert-barite unit from North Pole, Western Australia, *Alcheringa* **5**: 161–181.

Cady, S. L., Farmer, J. D., Grotzinger, J. P., Schopf, J. W., and Steele, A., 2003, Morphological biosignatures and the search for life on Mars, *Astrobiology* **3**: 351–368.

Grey, K., Hickman, A. H., Hofmann, H. J., Van Kranendonk, M. J., and Williams, I., 1999, *Pilbara Archaean stromatolite excursion field guide*, Western Australia Geological Survey Record.

Grotzinger, J. P., and Knoll, A. H., 1999, Stromatolites in Precambrian carbonates; evolutionary mileposts or environmental dipsticks? *Annual Reviews of Earth and Planetary Science Letters* **27**: 313–358.

Grotzinger, J. P., and Rothman, D. H., 1996, An abiotic model for stomatolite morphogenesis, *Nature* **383**: 423–425.

Hofmann, H. J., 2000, Archaean stromatolites as microbial archives. In: Riding, R. E., and Awramik, S. M. (Eds.) *Microbial Sediments*, Springer, Berlin.

Hofmann, H. J., 2004, Archean microfossils and abiomorphs, *Astrobiology* **4**: 135–136.

Lowe, D. R., 1994, Abiological origin of described stromatolites older than 3.2 Ga, *Geology* **22**: 387–390.

McLoughlin, N., 2006, *Earth's Earliest Biosphere: Western Australia*, D.Phil. thesis, Oxford University.

McLoughlin, N., Brasier, M. D., Wacey, D., Green, O. R., and Perry, R. S., 2007, On biogenicity criteria for endolithic microborings on early Earth and beyond, *Astrobiology* **7**: 10–26.

Pope, M. C., and Grotzinger, J. P., 2000, Controls on fabric development and morphology of tufas and stromatolites, uppermost Pethei group 1.8 Ga, Great Slave Lake, NW Canada. In: *Carbonate sedimentation and diagenesis in the evolving Precambrian world, SEPM Special Publication* **67**: 103–121.

Popper, K. R., 1959, *The Logic of Scientific Discovery,* Hutchinson, London, 480 p.

Schopf, J. W., and Walter, M. R., 1983, Archean microfossils: new evidence of ancient microbes. In: Schopf, J. W. (Ed.) *Earth's Earliest Biosphere, Its Origin and Evolution*, Princeton University Press, Princeton, NJ, pp. 214–239.

Semikhatov, M. A., Gebelein, C. D., Cloud, P., Awramik, S. M., and Benmore, W. C., 1979, Stromatolite morphogenesis: progress and problems, *Canadian Journal of Earth Sciences* **16**: 992–1015.

Seong-Joo, L., and Golubic, S., 1999, Microfossil populations in the context of syn-sedimentary micrite deposition and acicular carbonate precipitation: Mesoproterozoic Gaoyuzhuang Formation, China, *Precambrian Research* **96**: 183–208.

Sugitani, K., Grey, K., Allwood, A., Nagaoka, T., Mimura, K., Minami, M., Marshall, C. P., Van Kranendonk, M. J. and Walter, M. R., 2007, Diverse microstructures from Archaean chert from the Mount Goldsworthy-Mount Grant area, Pilbara Craton, Western Australia: microfossils, dubiofossils, or pseudofossils? *Precambrian Research* **158**: 228–262.

Turner, E. C., James, N. P., and Narbonne, G. M., 2000, Taphonomic control on microstructure in early Neoproterozoic reefal stromatolites and thrombolites, *Palaios* **15**: 87–111.

Walter, M. R., 1976, *Stromatolites* (Ed.), Elsevier, Amsterdam, 790 p.

Westall, F., and Folk, R. L., 2003, Exogenous carbonaceous microstructures in Early Archaean cherts and BIFs from the Isua Greenstone belt: implications for the search for life in ancient rocks, *Precambrian Research* **126**: 313–330.

# Chapter 4
# Fulfilling the Criteria for Early Life on Earth

## 4.1 Introduction

The 'burden of proof' needed for the demonstration of the earliest cellular life is very great indeed. Compelling proof requires the demonstration of multiple, *in situ* and mutually supporting lines of evidence to fulfil the aforementioned criteria: for a well-constrained age and context; evidence for a morphology unique to biology; and more than a single line of geochemical evidence for metabolic cycling; *together with* falsification of the null hypothesis of plausible non-biological origins (see for example Brasieret al., 2004, 2005, 2006 and references therein). This section of the book points the reader to the best areas of the world in which to search for signs of early life and the types of rocks to look for, before Chapter 5 discusses the techniques to use to analyse these rocks.

## 4.2 Where to Look? – Archean Cratons

The search for the earliest life on Earth relies, of course, on finding ancient rocks where biological signals may still be preserved. Unfortunately the majority of Earth's earliest rocks are stretched, squeezed or melted beyond recognition, providing obvious limitations to the fidelity of any reconstruction we may create from them. Although there are numerous examples of Archean (>2,500 Ma) rocks exposed at the surface of the earth at the present day (Fig. A6), there are only two places on Earth where one can find well exposed, unambiguous sedimentary rocks of low metamorphic grade and well constrained *early* Archean age. These two areas are the Pilbara craton of Western Australia, where the oldest sedimentary rocks are about 3,515 Ma, and the Barberton area of the Kaapvaal craton of southern Africa, where the oldest sedimentary rocks are about 3,470 Ma (Fig. A6). Indeed, the similarity of the rock records from these two ancient areas has led many to believe that they were once part of a single Archean 'super-craton' (termed Vaalbara; Zegers et al., 1998). South-west Greenland contains rocks that may be as old as ~3,800 Ma and are arguably sedimentary in origin (Fig. A6). These have been metamorphosed to a higher degree which makes any discussion of life signatures particularly controversial.

D. Wacey, *Early Life on Earth: A Practical Guide*,

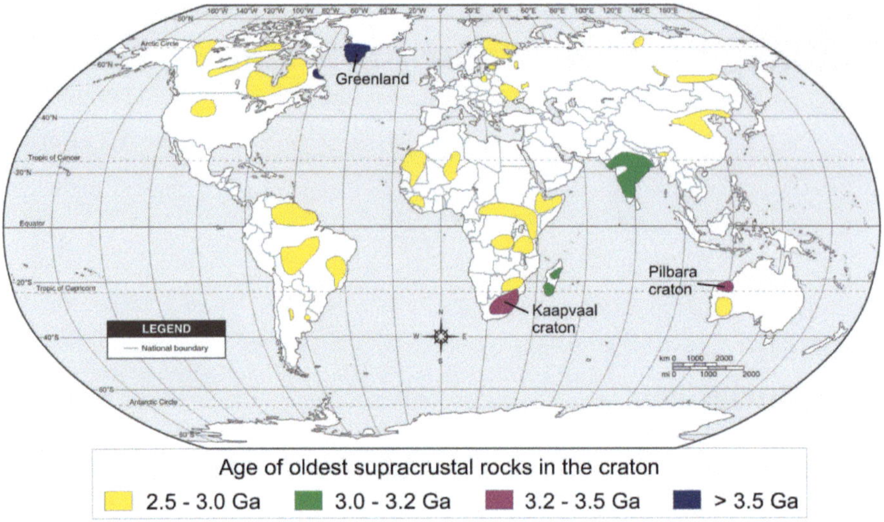


**Fig. A6** Remains of Archean cratons currently exposed at the Earth's surface. The earliest putative evidence for life on Earth comes from the Isua and Akilia regions of Greenland, the Barberton greenstone belt of the Kaapvaal craton, South Africa, and numerous greenstone belts within the Pilbara craton, Western Australia

For this reason, most of the illustrated putative examples of life that follow in Part B of this book are drawn from the Pilbara and Barberton rock units. A summary of the geology of the Pilbara, Barberton and south-west Greenland regions is given here to enable the context of claims of >3,000 Ma life to be understood. This book only deals with the particularly controversial time between ~3,800 and ~3,000 Ma. By later Archean (3,000–2,500 Ma) times, examples of microfossils and biochemical signatures are more widespread and more generally accepted. I will, however, use some comparative examples from the later Archean (e.g., Belingwe greenstone belt (Zimbabwe), Campbellrand, Pongola and Transvaal Supergroups (South Africa), and the Hamersley Basin (Western Australia)), plus Proterozoic and Phanerozoic analogues.

### *4.2.1 Geology of the Pilbara Craton*

The Pilbara craton of Western Australia (Fig. A7) is composed of three ancient granite greenstone terranes; East Pilbara, West Pilbara and Kurrana. The oldest rocks are exposed in the East Pilbara Terrane, the ancient core of the craton, where 3,655–2,850 Ma granitic bodies are interspersed with numerous volcano-sedimentary greenstone belts (~3,530–3,170 Ma) termed the Pilbara Supergroup. This Supergroup (Fig. A8) contains thick sequences of basalts and ultramafic lavas, with more minor felsic volcanics and volcaniclastics, numerous thin cherts and occasional sandstone and shale. It is overlain by the De Grey Supergroup, a 3,020–2,930 Ma largely clastic sedimentary sequence.

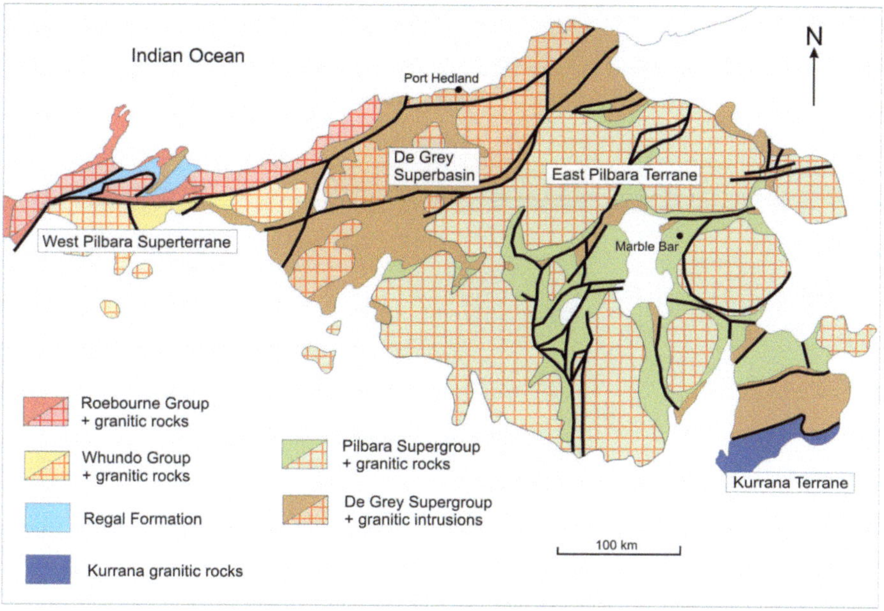


**Fig. A7** Geological map of the Pilbara Craton in Western Australia (Modified from Van Kranendonk et al., 2007). Studies of early life have been concentrated in numerous greenstone belts in the East Pilbara Terrane that contain Pilbara Supergroup rocks. There have also been recent studies in the West Pilbara Superterrane

The Pilbara Supergroup contains four unconformity bound stratigraphic intervals (Groups) – the ~3,520–3,427 Ma Warrawoona Group, ~3,350–3,315 Ma Kelly Group, ~3,270–3,230 Ma Sulphur Springs Group and the >3,000 Ma Soanesville Group (Fig. A8). The dips of each of the Groups gradually decrease with time, suggesting unconformable deposition on top of one another as thickening wedges adjacent to the growing granitic bodies. The lowermost Warrawoona Group consists mostly of mafic volcanic rocks of the Double Bar Formation, Table Top Formation, North Star Basalt, Mount Ada Basalt, and Apex Basalt. These volcanic rocks are intercalated with thin chert horizons and felsic volcanics of the 3,515–3,500 Ma Coucal Formation, the 3,472–3,465 Ma Duffer Formation, the ~3,460 Ma Marble Bar and Apex cherts (Figs. A9 and A10) and the 3,458–3,427 Ma Panorama Formation (Van Kranendonk et al., 2006). The 3,490 Ma Dresser Formation (Fig. A9) is a unique, geographically restricted unit, occurring only in the ~25 km diameter North Pole Dome. It consists of bedded chert, sulphate, carbonate and jasper, together with pillow basalt. The 3,426–3,350 Ma Strelley Pool Formation (Figs. A10 and A11) separates the Warrawoona Group from the overlying Kelly Group. The Kelly Group comprises, the 3,350–3,325 Ma Euro Basalt (Fig. A11), the 3,325–3,315 Ma Wyman Formation and the Charteris Basalt. This Group, in turn, is unconformably overlain by the Sulphur Springs Group comprising basal sandstone and felsic volcaniclastic rocks of the ~3,270–3,250 Ma Leilera Formation, komatiitic basalt of the ~3,250 Ma Kunagunarrina Formation, and felsic volcanic, epiclastic and

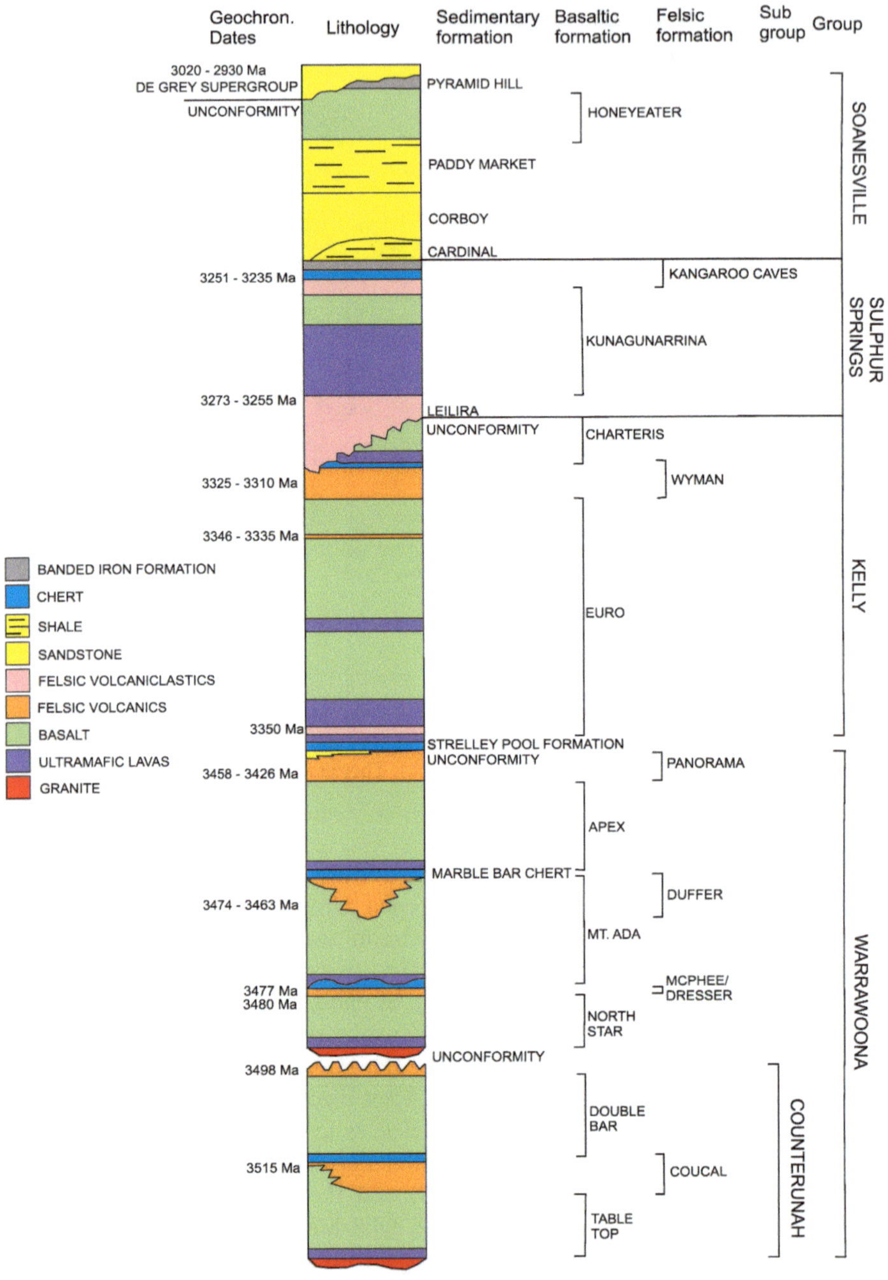


**Fig. A8** Stratigraphy of the Pilbara Supergroup in the East Pilbara Terrane of Western Australia (Modified from Van Kranendonk et al., 2007). Putative signs of life have been described from: Dresser Formation (Fig. A9 and pages 135–150); cherts within the Mt. Ada Basalt (pages 151–154), Apex Basalt (Fig. A10 and pages 155–159) and Panorama Formation (pages 175–179); chert and sandstone within the Strelley Pool Formation (Figs. A10, A11 and pages 181–197); pillow basalt and hyaloclastite within the Euro Basalt (Fig. A11 and pages 209–213); and chert and sulphide from the Kangaroo Caves Formation (pages 221–227)

**Fig. A9** Field photographs of important areas in the East Pilbara for studies of early life. *Top*, the ~3,490 Ma Dresser Formation in the North Pole Dome. Putative stromatolites, microfossils and biologically mediated pyrite have been described from this locality. *Bottom*, the ~ 3,465 Ma Marble Bar Chert at Marble Bar. This black, white and red banded and brecciated chert is an important unit for investigating the formation mechanisms of Archean cherts and the environmental conditions for earliest life

siliciclastic rocks of the ~3,240 Ma Kangaroo Caves Formation. The uppermost Soanesville Group is largely made up of clastic sedimentary rocks of the Cardinal, Corboy and Paddy Market Formations, with minor basalt (Honeyeater Basalt) and banded iron formation (Pyramid Hill Formation). Although as yet undated, the Soanesville Group is thought to have been deposited very soon after the Sulphur Springs Group at around 3,230–3,170 Ma.

In the West Pilbara Superterrane, the stratigraphy is somewhat different. Here, there are three separate stratigraphic successions occurring in three different tectonically separated terranes (~3,270 Ma Karratha, ~3,200 Ma Regal and 3,120 Ma Sholl terranes). The oldest supracrustal rocks in the West Pilbara belong to the ~3,270–3,250 Ma Roebourne Group that is restricted to the Karratha terrane.

**Fig. A10** *Top*, the ~3,460 Ma Apex Basalt at Chinaman Creek, Marble Bar greenstone belt. Putative microfossils have been described from the hydrothermal silica vein (*outlined*) intruding a stratiform chert unit. *Bottom*, the ~3,400 Ma Strelley Pool Formation at the Trendall locality, North Pole greenstone belt. Putative biological stromatolites have been described from this locality

This succession is dominated by ultramafic to mafic volcanics (Ruth Well Formation) with subordinate overlying clastic meta-sedimentary rocks (Nickol River Formation). The Roebourne Group is interpreted to have formed on East Pilbara basement during

**Fig. A11** *Top*, Strelley Pool, East Strelley greenstone belt. Here the ~3,400 Ma Strelley Pool Formation is conformably overlain by the ~3,350 Ma Euro Basalt. Putative stromatolites, microborings and biominerals have been described from this area. *Bottom*, Sulphur Springs Creek, East Strelley greenstone belt. Here chert ridges from the ~3,515 Ma Coonterunah Subgroup (*outlined*) produce an angular unconformity with the overlying Strelley Pool Formation. This is thought to represent Earth's oldest exposed land surface rifting of the East Pilbara Terrane, after which the East and West Pilbara evolved in different tectonic environments (Hickman, 2004). In the Regal Terrane the oldest rocks are mid-ocean-ridge-type basalts of the Regal Formation that are undated, but older than 3,160 Ma. These are overlain unconformably by the ~3,020 Ma Cleaverville Formation. This comprises banded iron-formation, ferruginous chert, grey–white and black chert, shale, siltstone and minor volcanogenic sedimentary rocks. The Cleaverville Formation also occurs in the Sholl terrane, unconformably overlying the dominantly basaltic 3,130–3,115 Ma Whundo Group.

### 4.2.2 *Geology of the Barberton Greenstone Belt, Kaapvaal Craton*

The Barberton greenstone belt is found in the eastern part of the Kaapvaal Craton in the Transvaal province of South Africa and in bordering Swaziland (Fig. A12). The supracrustal (i.e., volcanic and sedimentary) rocks in this area are assigned to the ~3,550–3,220 Ma Swaziland Supergroup (Fig. A13). This is subdivided into three stratigraphic groups; a lower, mostly volcanic succession (~3,550–3,300 Ma Onverwacht

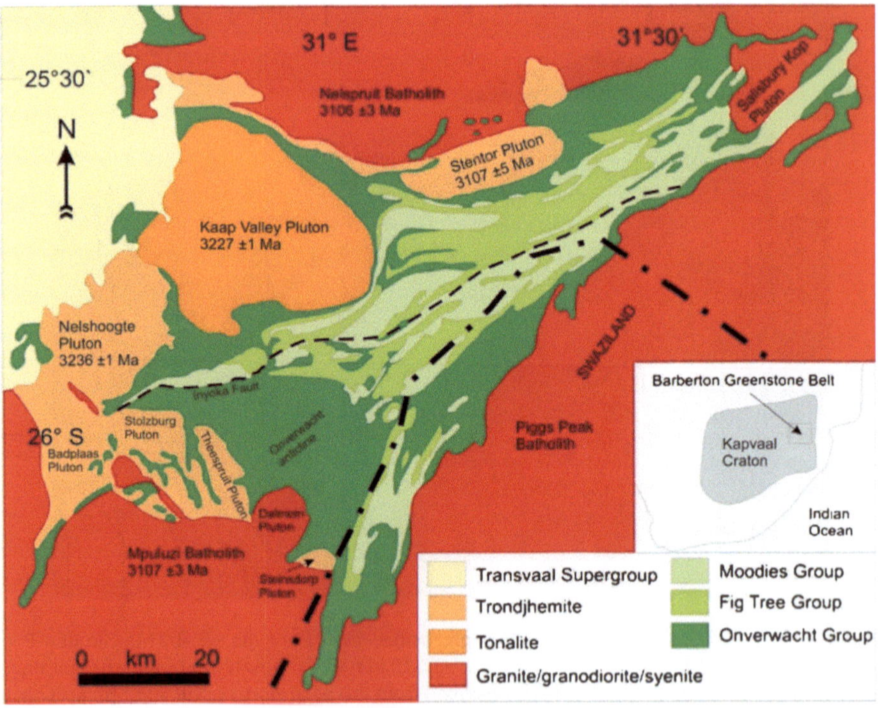


**Fig. A12** Simplified geological map of the Barberton greenstone belt, South Africa and Swaziland. Evidence of early life on Earth has been reported from the Onverwacht (see pages 161–174 and 199–208), Fig Tree (see pages 215–219) and Moodies Groups (see pages 229–234)

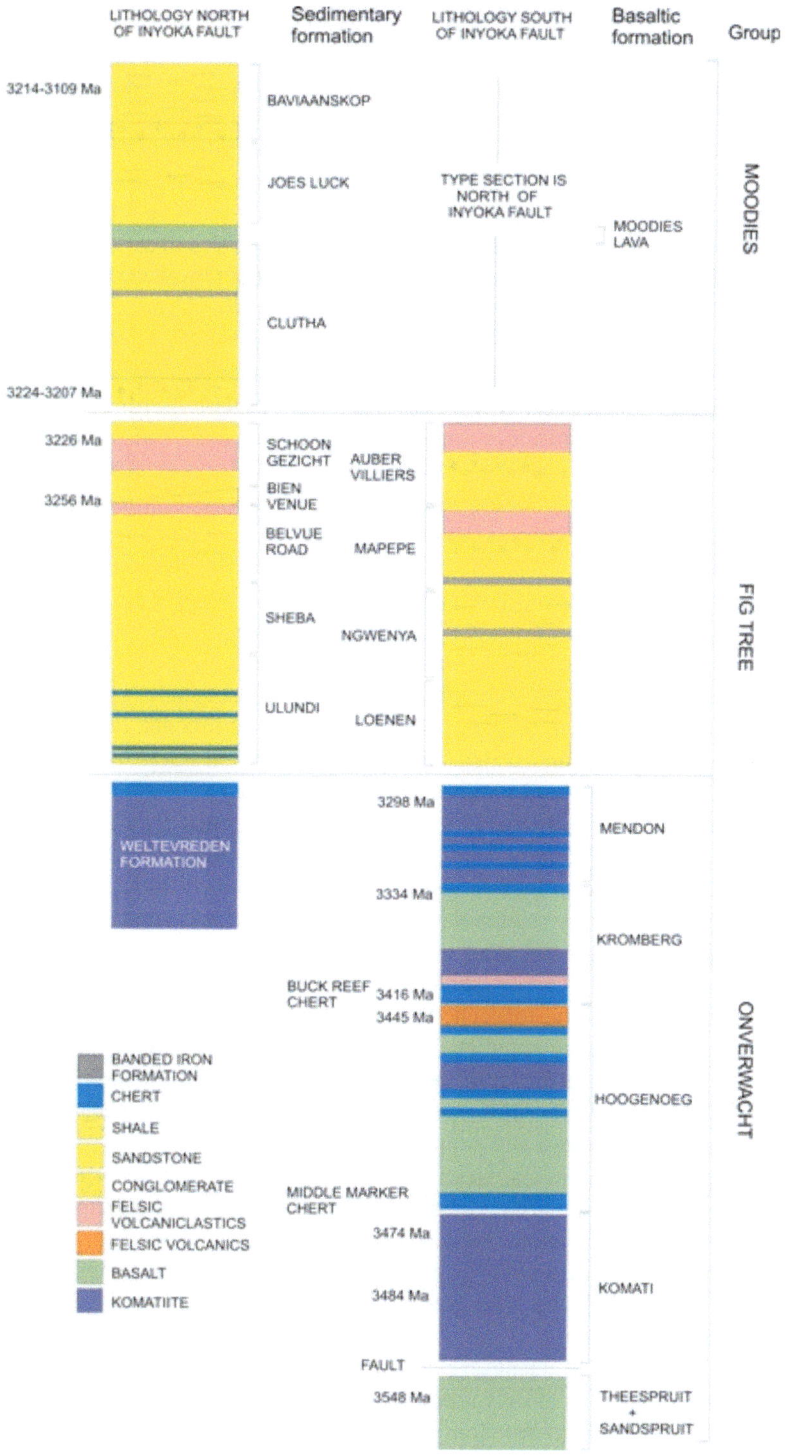


**Fig. A13** Stratigraphy of the Barberton greenstone belt, southern Africa (see text on pages 64–67 for details)

Group) followed by two mainly clastic successions (~3,260–3,225 Ma Fig Tree Group and ~3,220 Ma Moodies Group). These are surrounded and intruded by plutonic trondhjemite-tonalite and granodiorite domes and sheets of 3,500–3,100 Ma.

The stratigraphy of the Swaziland Supergroup is somewhat variable from north to south across the greenstone belt. The Inyoka fault zone (Fig. A12) serves as a useful marker to separate the northern and southern sections and appears to represent a tectono-stratigraphic boundary. The type section of the Onverwacht Group lies to the south of the Inyoka fault where it is divided into the Theespruit, Sandspruit, Komati, Hoogenoeg, Kromberg and Mendon Formations, a series of komatiitic and tholeiitic basaltic rocks (e.g., Fig. A14) inter-bedded with thin sedimentary units of silicified ash and black chert (Figs. A14 and A15), together with somewhat rarer felsic volcaniclastic and intrusive rocks. The Theespruit and Sandspruit Formations are only seen in faulted contact with the rest of the group, and are more highly metamorphosed, whilst the Komati Formation is composed entirely of komatiitic lavas, both unlikely targets for signs of life. In contrast, putative signs of life have been reported from cherts and pillow basalts from the Hoogenoeg and Kromberg Formations (see pages 161–174 and 199–208).

The Fig Tree Group in the north of the area is composed of turbiditic greywacke, shale and felsic volcanics, whilst in the south it is much more variable, apparently having been deposited in shallower water. Here it comprises shale, greywacke, minor conglomerate, fine grained tuffaceous and ferruginous sediments, chert, banded iron formation and felsic volcaniclastics, and is at least 1,200 m thick. The contact with the underlying Onverwacht Group is sharp but interpreted to be conformable. To the north of the Inyoka Fault, the Fig Tree Group is divided into five formations (Fig. A13): Ulundi (shale and banded chert); Sheba (sandstone and shale interbeds); Belvue Road (shale and minor sandstone); Bien Venue (volcaniclastic sediment); and Schoongezicht (shale, volcaniclastic sandstone and conglomerate). To the south of the Inyoka Fault, four different Formations are recognised (Fig. A13): Loenen (sandstone and shale); Ngwena (shale, fine sandstone and banded iron formation); Mapepe (conglomerate, sandstone, chert, shale, siltstone); and Auber Villiers (sandstone, conglomerate). Correlation of these formations across the fault zone has proved difficult. Putative signs of life have been described from the Sheba Formation (see pages 215–219), whilst spherules interpreted to be of meteorite impact origin, occur at the base of both northern and southern Fig Tree successions.

The Moodies Group lies unconformably above the older sediments and volcanics. It is made up of shallow marine, deltaic and sub-aerially deposited sandstones with lesser amounts of conglomerate, siltstone and shale, and rare volcanic units. The group appears to have been deposited in at least two separate basins which now form structurally separate blocks. The type section is to the north of the Inyoka fault, where the Moodies Group is up to 3,700 m thick (Fig. A13), divided into the Clutha, Joe's Luck and Baviaanskop Formations, whilst to the south of this fault it has a maximum thickness of only around 1,000 m. The southern block may represent time-equivalent deposition but from a different source area to the lowermost units of the northern block. However, unambiguous correlation between Moodies rocks to the north and south of the Inyoka fault has yet to be achieved. Nevertheless, this

**Fig. A14** Important localities for the study of early life in the Barberton greenstone belt, southern Africa. *Top*, typical pillow lavas of the Kromberg Formation containing microtubular structures which are claimed to be biological. *Bottom*, the Komati River cutting through the eastern limb of the Onverwacht Anticline. Cherts from the upper Hoogenoeg and lower Kromberg Formations outcrop in this river section and contain putative evidence for early life in the form of microfossils and microbial mats (Photographs are courtesy of Nicola McLoughlin and Martin Brasier)

**Fig. A15** Two views of a typical chert unit in the Barberton greenstone belt. This chert unit is from the middle of the Hoogenoeg Formation and can be seen for over 40 km in both the west and east limbs of the Onverwacht Anticline. It contains carbonaceous particles which may be biological in origin (Photographs are courtesy of Martin Brasier)

succession represents Earth's oldest preserved relatively un-metamorphosed sequence of siliciclastic sediments and putative signs of life have been reported from sandstones in the lower part of this Group (see pages 229–234).

As with the Pilbara, the Barberton is structurally very complex with most rocks dipping almost vertically with tight folds, widespread faulting and structural repetition. There has also been widespread post-depositional alteration of these rocks, although metamorphism has not been too intense (maximum of greenschist to amphibolite facies) allowing some primary mineralogical and textural assemblages to be preserved. Detailed geological mapping is therefore imperative so that the age and context of any units containing signs of life can be understood.

### 4.2.3 *Geology of South-West Greenland*

The Isua greenstone belt and Akilia Island in south-west Greenland (Fig. A16) contain the oldest intact supracrustal (i.e., volcanic and sedimentary) rocks on our planet. These have a minimum age of ~3,700 Ma in Isua (Moorbath et al., 1973) and could be as old as ~3,850 Ma on Akilia Island (Nutman et al., 1997). Unfortunately these rocks have been subjected to intense metamorphism, so they are difficult to decode. There has been much debate and controversy surrounding the value of these Greenland successions to provide the oldest possible evidence of life (see for example a recent review by Whitehouse and Fedo, 2007).

The first problem facing researchers in Greenland is establishing whether or not the rocks are sedimentary in origin. The tiny island of Akilia (Fig. A16), just off the coast of south-western Greenland has provided the backdrop to one such investigation. The majority of rocks on Akilia are highly unsuitable for searching for early life. Akilia is mostly composed of igneous mafic and ultramafic rocks, metamorphosed to at least amphibolite facies, together with quartz-feldspar gneiss. However, there is one small outcrop of banded quartz-pyroxene rock, only a few tens of square meters in total area, that may provide the first evidence for sedimentary rocks on Earth. This outcrop was initially interpreted as a metamorphosed sedimentary banded iron formation (BIF) (e.g., Nutman et al., 1996) and much excitement arose from the discovery of carbonaceous material within this outcrop (Mojzsis et al., 1996). However, a second team of geologists (Fedo and Whitehouse, 2002) examined the Akilia site and determined that the outcrop exhibits nothing to indicate an original sedimentary environment. A re-examination of the field relationships of the outcrop showed the banding to comprise discontinuous boudinage tails caused by multiple, intense deformation events. Furthermore, detailed geochemical data (major elements, trace elements and rare earth elements) from this team pointed to an ultramafic igneous protolith (like much of the rest of Akilia), not a sedimentary one. Crucially, an igneous protolith for this rock means that any carbon found within it would have little or no biological relevance.

This story took a further twist when Dauphas et al. (2004) came out in support of the sedimentary BIF origin for this deposit. They used a new approach involving iron isotopes and found that the Akilia rocks were enriched in the heavy iron isotope when compared to igneous rocks. This, they argued, was consistent with the transport, oxidation and precipitation of ferrous iron from hydrothermal vents and thus consistent with these rocks indeed being the oldest water lain sedimentary deposit

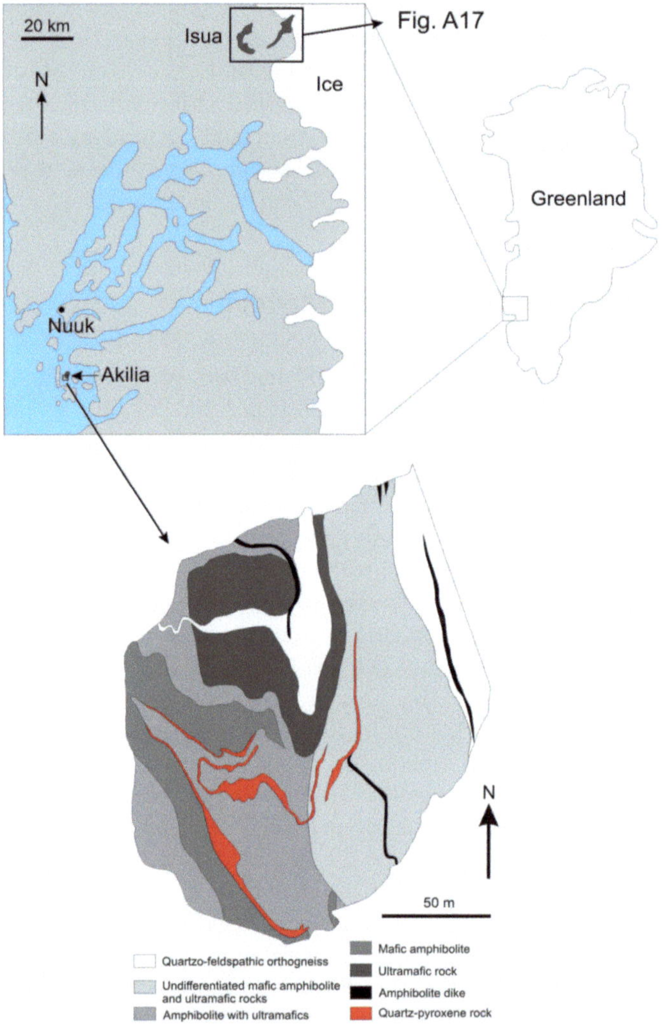


**Fig. A16** Location of the Isua greenstone belt and Akilia Island in south-west Greenland. Enlargement shows the small area of Akilia Island containing the controversial banded quartz-pyroxene rock, interpreted by some to be Earth's oldest sedimentary rock

preserved on Earth. This argument is ongoing, with new mapping, geochronology and geochemistry (e.g., Manning et al., 2006) increasing our knowledge of this tiny island. However, until the controversy is satisfactorily resolved, any putative reports of early life emanating from Akilia must be viewed with extreme caution.

A more promising location to search for ancient biosignatures in Greenland appears to be the Isua greenstone belt (ISB), in particular a low strain domain in the north east of the belt (Fig. A17). Here, although almost all the rocks are deformed and altered by metasomatism, it is still possible to see transition stages between the original protoliths and the dominantly schistose rocks we see today. It is possible to clearly

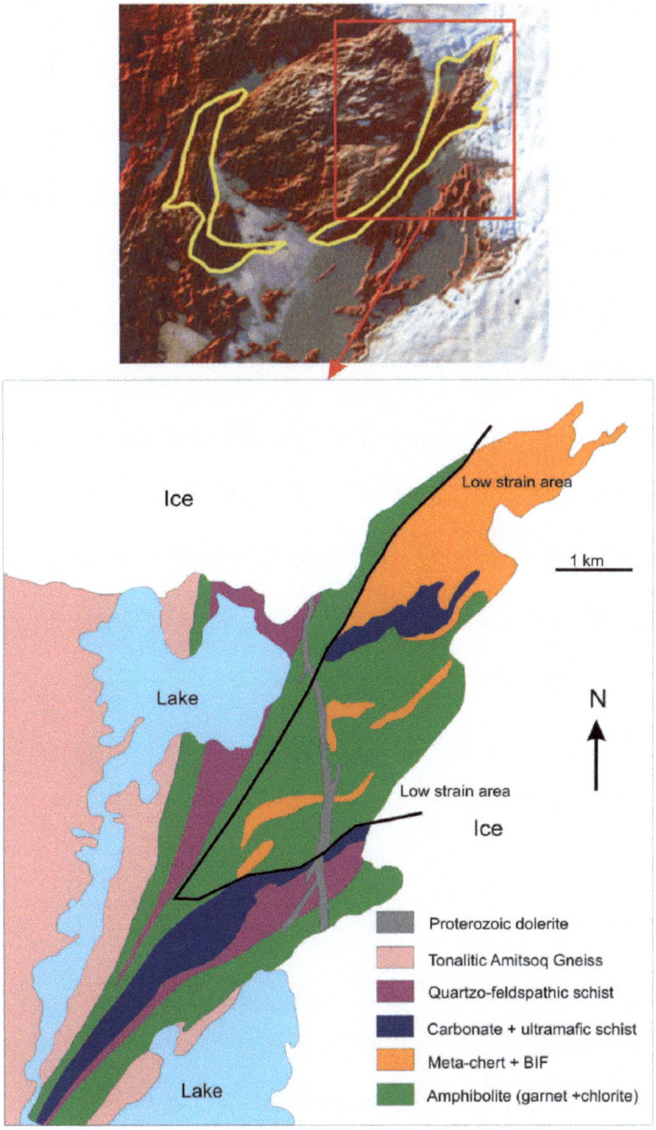


**Fig. A17** Location of the Isua greenstone belt (ISB) in south-west Greenland. *Top*, ASTER satellite photograph of part of south-west Greenland with the ISB outlined in yellow. *Bottom*, enlargement of boxed area, showing the low strain area (north-east of *black line*) of the ISB where the most promising rocks in the search for Earth's earliest life may be found

identify examples of Earth's oldest pillow lavas (Fig. A18), banded iron formations (Fig. A19) and clastic rocks including conglomerates and turbidite sequences (Fig. A19). Claims for early Archean life have come from both iron carbonate-rich sedimentary rocks (Mojzsis et al., 1996) and (meta)turbidites (Rosing, 1999) from this area (Fig. A20 bottom).

**Fig. A18** Field photographs of areas in south-west Greenland relevant for studies of early life. *Top*, typical Greenland terrain. *Bottom*, earth's oldest preserved pillow lavas, similar to those from the Barberton greenstone belt and Pilbara craton that contain putative Archean life (Photographs courtesy of Stephen Moorbath)

The second problem facing researchers in Greenland, both with Isua and Akilia rocks, is the high degree (and often multiple episodes) of metamorphism that they have experienced. Any fossilized morphological remains of Earth's earliest biosphere, if any ever existed at that time, will have been destroyed by heat and pressure.

**Fig. A19** Field photographs of areas in south-west Greenland relevant for studies of early life. *Top*, Earth's oldest conglomerate deposit. Conglomerates, having an undoubted sedimentary origin and wide diversity of rock types contained within them as clasts, are ideal targets to search for early life. Scale: stick is 20 cm long. *Bottom*, typical Greenland banded iron formation (Photographs courtesy of Stephen Moorbath)

This means that we must rely solely on chemical signatures within these rocks that may give clues to the former existence of life. Chemical signatures thus far reported from Greenland rocks are dominated by carbon isotope fractionations. Some of these signatures have been consistent with ancient metabolic activity and biological processing (e.g., Rosing, 1999; see pages 129–133) but other lines of evidence are now required to make these claims compelling.

The debate surrounding the ~3,700–3,850 Ma rocks from Greenland is ongoing and, despite the great challenges associated with the Akilia and Isua areas, these

**Fig. A20** Two rock units from which claims of early life have been documented. *Top*, the highly controversial banded quartz-pyroxene rock from Akilia Island. This has been interpreted as either a metamorphosed sedimentary banded iron formation, or a metasomatized ultramafic unit. *Bottom*, a banded rock interpreted as part of a sedimentary turbidite sequence from the Isua greenstone belt. Carbon isotope signatures of $\delta^{13}C = -19‰$ PDB from graphite globules within this rock have been claimed to be evidence of early life (Photographs courtesy of Stephen Moorbath)

rocks are unique in representing the period in Earth history where geological processes as we know them today may first be recognizable and where conditions may have first become tolerable for life. The fact that unambiguous signals for life have not been forthcoming, as yet, should not deter further detailed study into this earliest window into potential life on Earth.

## 4.3 Typical Rocks Found in the Early Archean That Could Host Life

### 4.3.1 Chert

A large proportion of the (meta)-sedimentary rocks in the Pilbara and Barberton regions are cherts. A chert is defined as a sedimentary rock composed largely (>75%) of microcrystalline silica in the form of microquartz or chalcedony (Folk and Weaver, 1952), but this definition reveals little about the processes that result in chert formation. The fine grain size of chert, combined with its common formation during early diagenesis in near surface environments, makes it an ideal lithology for the preservation of morphological and chemical remains of life. Historically, chert has been the *key* lithology in the search for early life on Earth.

The process of chert formation in the Archean was very different to that in the Phanerozoic and represents a prime example of the breakdown of the 'Principle of Uniformity' (see Section 2.4). In the Phanerozoic the oceanic silica budget is largely controlled by microorganisms (e.g., diatoms and radiolaria) that secrete siliceous skeletons. These organisms keep the concentration of silica in the ocean at around 1 part per million (ppm) or less. In contrast, there is no evidence for silica secreting organisms in the Archean and oceanic silica concentrations have been estimated at about 60 ppm (Siever, 1992). In the Archean, factors such as temperature, pH and exchange with silicate minerals likely controlled the precipitation of silica. This is not to say that biology did not have some role to play in silica precipitation in the Archean – this remains to be fully tested.

Most simply, chert can form as a primary chemical precipitate from waters supersaturated with respect to silica. This mechanism is, however, rather rare and the majority of Archean chert appears to have formed by secondary silicification of other deposits such as volcanic ash, volcaniclastics or, more rarely in the Archean, silt, mud and carbonate. This secondary process can take place either early in the history of the deposit by low temperature silica saturated seawater, or later within the sediment column by circulating hydrothermal fluids (Lowe, 1999).

Chert is often the easiest type of rock to spot in the field as it is resistant to weathering and forms long narrow ridges within greenstone belts. Unfortunately this ease of recognition does not extend to chert interpretation. Many types of sediments and igneous rocks can be silicified to form chert and only some of these primary lithologies are suitable for life. In addition, chert may result from more than one episode of silicification, so detailed observations under the microscope are needed to decode the history of the chert and place any putative life signals in the correct time frame.

As a first order observation in the field, the colour of chert may give some clue to its origin. Chert derived from felsic volcanic ash tends to be pale to medium grey, or even yellowish grey, and may contain some preserved quartz phenocrysts. In contrast, chert derived from komatiitic ash tends to be pale green, grey green or bluish grey in colour (Fig. A21a) and accretionary lapilli are often widespread.

**Fig. A21** The variety of Archean chert. (**a**) Pale green chert representing silicified komatiitic ash, intruded by later black chert from nearby hydrothermal vein, ~3,400 Ma Strelley Pool Formation in the East Strelley greenstone belt. (**b**) Silicified volcaniclastic sediment capping a thick basalt unit within the ~3,350 Ma Euro Basalt, East Strelley greenstone belt. Layers of black chert alternate with grey chert containing volcaniclastic debris. (**c**) Typical black, white and grey banded chert with later stage fractures and chert veins. This shows how multiple episodes of silicification need to be untangled to decode the age of some putative signs of life. Sample from the Archean Biosphere Drilling Project through the Strelley Pool Formation. (**d**) Banded ferruginous chert from the ~3,490 Ma Dresser Formation in the North Pole Dome. (**e**) Banded grey, white and red (jaspilitic) chert from the Marble Bar Chert at Marble Bar (scale as for d). (**f**) Black hydrothermal chert intruding the ~3,400 Ma Strelley Pool sandstone (palaeosurface is *up* on the photograph). This chert clearly formed after solidification of the sandstone as evidenced by the fact that angular clasts of sandstone have been ripped up and incorporated in the black chert (to the *right* of the pen). Small fingers of chert also penetrate horizontal weaknesses in the sandstone (Photographs by the author)

Chert derived from basaltic debris commonly caps basaltic units and is likely black and often carbonaceous (Fig. A21b). Laminated black/grey chert (Fig. A21c) most probably represents mixtures of ash, carbonaceous matter and orthochemical minerals.

Banded ferruginous chert (white and rusty brown bands in outcrop, Fig. A21d) may have a basaltic component, with additional chemical precipitation of siderite

($FeCO_3$) and pure silica enhancing the banding. Silicified evaporites can often be recognised by pseudomorphs of primary evaporitic minerals such as gypsum ($CaSO_4$) or nahcolite ($NaHCO_3$), although the environments where evaporites formed in the Archean may have been rather different to today due to differing ocean chemistry. Red chert (jasper, Fig. A21e) has been argued to be a primary iron oxide/silica precipitate. Silicified carbonates bring an extra dimension to the problem because the carbonate may itself have been a replacement of a primary mineralogy. In most cases, geochemical analysis (bulk mineralogy and trace elements) in the laboratory are necessary to confirm the chert protolith and deduce its environment of deposition.

The *timing* of silicification has become vitally important in studies of early life. For example, if silicification of the primary sediments was very early, prior to sediment compaction and any metamorphism, then any putative biological structure preserved within the silica can be confidently given an Archean age. If, however, silicification was late, well after sediment compaction (Fig. A21f), then the age of any associated biological structures is much less certain. Careful field and petrographic observations will, in most (but maybe not all!) cases, be able to distinguish between these two options.

The most widely studied type of chert in the Archean is that with a significant carbon content (carbonaceous chert) since carbon may mean life; these are mostly represented by dark grey, black and black-and-white banded chert varieties. These can be massive, planer laminated or wavy laminated. Many of these cherts show evidence of their protolith being deposited in shallow water current or wave active regimes (e.g., Buck Reef Chert, Figs. B54 and B55; Tice and Lowe, 2004), but some are clearly deposited under much quieter, deeper water conditions.

In the Pilbara, chert units are widespread, the oldest dating back to ~3,515 Ma within the Coonterunah Subgroup (Fig. A11). Important chert units for the study of early life occur within the ~3,490 Ma Dresser Formation, ~3,460 Ma Apex Basalt, ~3,450 Ma Panarama Formation, ~3,400 Ma Strelley Pool Formation, ~3,240 Ma Sulphur Springs Group, and the ~3,000 Ma Gorge Creek Group.

Cherts of the **Dresser Formation** are complex (Figs. A9 and A21d); they occur as a series of bedded chert-barite-carbonate units with significant hydrothermal input evidenced by swarms of siliceous feeder veins emplaced along growth faults (Van Kranendonk, 2006). It has been suggested that the protolith of this chert was carbonate based on evidence for pseudomorphs of carbonate rhombs and oncolitic structures; indeed primary carbonate (ankerite, $Ca(Fe,Mg,Mn)(CO_3)_2$) is preserved at one locality (Garcia-Ruiz et al., 2003). Hence, the chert-barite-carbonate succession seen in the field today may represent preferential (almost contemporaneous) silicification of certain layers of an original carbonate sediment formed in shallow seawater. The barite ($BaSO_4$) is thought to be primary, brought into the system by the same hydrothermal fluids that caused silicification during faulting and early basin development (Van Kranendonk, 2006). Putative biological stromatolites and biominerals from the stratigraphically lowermost chert-barite unit are critically reviewed on pages 135–149.

The '**Apex chert**' generally refers to the thickest and lowermost of a series of thin cherts within the **Apex Basalt**. Bedded cherts in this Formation are typically silicified felsic tuffs occurring at the break between basaltic volcanic cycles.

The 'Apex chert' itself is underlain and intruded by a series of black chert veins (also sometimes referred to as chert dykes). These chert veins appear to have been emplaced along growth faults during hydrothermal activity that was approximately synchronous to deposition of the bedded chert (Van Kranendonk, 2006). Here, silicification has been interpreted as occurring in a submarine hot-spring environment (Van Kranendonk and Pirajno, 2004). Controversial filamentous 'microfossils' were described from the bedded 'Apex chert' (Schopf, 1993) but were later shown to instead occur in one of the hydrothermal black chert veins (Brasier et al., 2002, 2005). These are critically reviewed on pages 12–15 and 155–159.

The '**Kitty's Gap Chert**' of the **Panorama Formation** has been interpreted as silicified near-shore volcaniclastic sediments (Westall et al., 2006). The source of the silica was interpreted to be partly hydrothermal and partly from devitrification of the glassy rhyolite protolith. Putative microbial colonies which may have played some part in the silicification process (Westall et al., 2006) are critically reviewed on pages 175–179.

Much of the **Strelley Pool Formation** is composed of chert. The protolith of the majority of this chert appears to be dolomite; this original carbonate is observed in numerous localities (e.g., Trendall Locality, Fig. A10) where only partial silicification has occurred. Trace element geochemistry has been used to infer a primary 'normal' marine origin (rather than hydrothermal) for the dolomite (Van Kranendonk et al., 2003). Silicification of the unit was, however, due to hydrothermal activity; black chert veins cut up through several hundred metres of underlying volcanic rocks and terminate towards the top of the Formation. This relationship indicates that silicification occurred shortly after deposition of the carbonate protolith and was likely caused by hydrothermal fluids driven by the heat associated with the eruption of the overlying basalts of the Euro Basalt (Van Kranendonk and Pirajno, 2004). Towards the top of the Formation, pale green and grey cherts are interpreted as having a volcanic ash protolith. This probably represents the onset of a period of volcanic activity that would go on to produce the overlying Euro Basalt. Putative biological stromatolites from the silicified carbonate unit are critically reviewed on pages 9–12, 184, and 191–194.

The protoliths of the marker chert unit within the **Sulphur Springs Group** have been interpreted as epiclastic sediments (Van Kranendonk, 2006). Silicification is again interpreted to be as a result of hydrothermal activity, evidenced by silica veins emplaced along growth faults, occurring almost contemporaneously with deposition of the sediments. Intrusion of the Strelley Monzogranite at ~3,240 Ma appears to have produced the necessary heat to drive the hydrothermal circulation in this area. Putative microfossils from these silicified sediments are critically reviewed on pages 222–227.

Finally, the red-white-black banded **Marble Bar Chert** (Figs. A9 and A21e), although not home to any claims for early life as yet, presents evidence for at least three phases of chert precipitation all linked to hydrothermal activity. Firstly, there is the jaspilitc (red) chert; this shows very fine planer laminations indicative of deposition in a quiet, deep water setting (Sugitani, 1992), possibly as a primary precipitate near to hydrothermal vents. The white chert appears to be a hydrothermal replacive phase, cutting across bedded jasper, but evidence of soft sediment deformation suggests that the two cherts are almost contemporaneous. In contrast, the black chert frequently occurs as branching veins cutting up through the bedded chert, and enclosing angular clasts of red-white banded chert. These discordant veins feed black chert bands that superficially appear concordant with the bedded red and white cherts.

In the Barberton, numerous thin chert units in the **Hooggenoeg**, **Kromberg** and lower **Mendon** Formations (e.g., Fig. A13) appear to have been formed almost syn-depositionally in a shallow water, platform setting (Lowe, 1999). Here, they have formed through the silicification of several different primary rocks types; volcaniclastic and pyroclastic deposits; terrigenous detrital layers; possible biological sediments; and chemical evaporitic deposits. In the upper parts of the Mendon Formation black, black-and-white banded, and banded ferruginous cherts represent silicified fine grained sediments deposited under low energy, deep water conditions. These were initially mixtures of tuffaceous, carbonaceous, calcareous (some sideritic) and locally primary siliceous sediments. Putative microfossils, trace fossils and chemical fossils have been reported from several of these chert units and are illustrated and critically reviewed on pages 161–173 and 199–208.

## *4.3.2 Pillow Basalt*

The search for life has recently been extended to pillow basalts (Fig. A22), associated hyaloclastite (glassy shards that have spalled off pillows during eruption), and glassy tuffs. Pillow basalts are not intuitively where one would look for life, since they are associated with high temperature volcanic activity. However, on closer examination, a mode of life within newly extruded sea floor pillow basalts would have many attractions for primitive microbes: a direct source of geothermal energy; a source of reductants, oxidants and biolimiting elements; and protection against harsh surface conditions including high UV radiation and the destructive effect of

**Fig. A22** Modern pillow basalts on the South Pacific seafloor (Image courtesy of NOAA)

rather large and frequent meteoritic impacts. In the Archean, the pillow basalt habitat was very much more widespread than today. In terms of thickness, pillow basalts can be seen to comprise over 99% of the stratiform succession in the greenstone belts of both the Barberton and Pilbara regions.

Potential evidence for life comes in the form of microborings from the glassy margins of pillow basalts and inter-pillow hyaloclastites. These are interpreted to have formed through biologically mediated corrosion of volcanic glass by endolithic microbes, and have been recently described from both the Barberton and Pilbara regions. In the Barberton, the microborings come from various levels within the Komati, Hooggenoeg and Kromberg formations of the Onverwacht Group, with the best developed examples from the 3,472–3,456 Ma upper Hooggenoeg Formation (Figs. B26 and B27; Furnes et al., 2004). Here, the microborings are mostly tubular in morphology, 1–10 μm in width and up to 200 μm in length. Segmentation is observed but this is likely merely due to the overgrowth of metamorphic minerals. Granular alteration is also observed in individual glass shards within inter-pillow hyaloclastite; this takes the form of individual and coalesced spherical bodies 1–4 μm in diameter. Biogenicity is further suggested by enrichments of carbon, and possibly also nitrogen and phosphorus within the microtubes, and by rather light bulk $\delta^{13}C$ values. However, the resolution of these analyses is rather poor and does not provide compelling evidence.

In the Pilbara, morphologically comparable microtubular structures occur within inter-pillow hyaloclastite from the 3,350 Ma Euro Basalt (Figs. A23 and B63;

**Fig. A23** Well preserved pillow basalts from the ~3,350 Ma Euro Basalt, Pilbara, Western Australia (Photograph by the author)

Banerjee et al., 2007). These are infilled with titanite ($CaTiSiO_5$) which has been dated directly using U-Pb systematics demonstrating that microtube formation took place prior to an Archean (~2,900 Ma) phase of metamorphism. In other words, these microtubes are unlikely to be younger contaminants. Biogenicity is inferred using similar evidence to the Barberton examples.

The identification of similar structures in basalts from *in situ* oceanic crust (ranging from <6 to ~180 Ma) and ophiolites of various ages lends supporting evidence to the biogenicity of the Archean examples. In Phanerozoic examples, it has been found that the biological alteration structures show several characteristic features that distinguish them from non-biological alteration (summarised from Staudigel et al., 2006):

- Bioalteration features often contain microbial cells, microbial DNA, carbon residues, and/or biologically distinct enrichments and depletions of elements.
- Bioalteration is never found completely enclosed in glass; it is always rooted on surfaces that are exposed to external water.
- Tubular and granular alteration locations on conjugate sides of a crack do not line up with one another. In contrast, physico-chemical alteration of basaltic glass (often termed palagonitisation) results in a number of mineral phases that replace the original glass; including smectite, phillipsite, chabazite and palagonite itself. These minerals tend to occur as bands, symmetrically arranged around fractures or coating vesicles.
- Tube and granule diameters are of micron to submicron scale, like microbes. Tubes tend to be larger than granules, yet both display log-normal size distributions, a common attribute in biological systems.
- Tubular alteration does not show flaring at the entry point or narrowing deeper inside the glass, as would be expected from non-biological dissolution.
- Some tubes show segmentation, in which the diameter of tubes varies regularly. This is highly suggestive of pulsed growth and/or the presence of several cells.
- Some tubes bifurcate, which can be explained by cell division.
- Some tubes show spirals that are extremely hard to generate non-biologically with the regularity observed. Spirals are common in biology and biologically produced materials.
- Granular alteration often forms hemispherical agglomerations of cavities, radiating out from a single point at a crack surface and producing the texture of a sponge. These agglomerations closely resemble the growth of microbial cultures on an agar dish, except that they are three-dimensional.

Although, inevitably and unfortunately, many of these characteristics are no longer preserved in the early Archean microtubes, the numerous morphological and chemical similarities between the Archean and more modern examples make these structures good candidates for being formed by some of Earth's oldest life forms.

Clues to possible microbial metabolisms come from modern in situ basaltic crust. Optimum growth conditions for the microbes appear to be 20–80°C and 0–300 m depth, but they have been observed to around 110°C and over 550 m, where water can circulate through networks of fractures within the volcanic pile. Both Archaea and Bacteria have been isolated from deep sea cores and it has been suggested that chemo(litho) autotropic iron and/or manganese oxidation are likely metabolic pathways.

### 4.3.3 Sandstone

Although the majority of claims for early Archean life come from cherts, these are a notoriously difficult rock type in which to decode post-depositional alteration. This means that it has proved extremely difficult to accurately determine the biogenicity and antiquity of any putative biological signals in chert, in turn leading to much controversy. In contrast, sandstones appear to provide a clearer window into the earliest biosphere. Well-preserved sandstones allow successive depositional and diagenetic stages to be identified within clasts, matrices, cements and veins, while the diverse composition of clasts allows the mapping of any preferential behaviour of biosignals (e.g., microborings). Sandstones are a 'strong' lithology and offer protection from later stress and strain, so that microscopic fossils/trace fossils are not easily destroyed by diagenetic processes. Indeed, sandstones may preserve the only relic clasts of lithologies long since eroded away. Sandstones clearly have the potential to preserve a wide range of morphological and geochemical signatures, but their rarity in the early Archean means they have only recently been studied in any detail.

Although sandstones and conglomerates have been recognised from ~3,700 Ma rocks in Greenland (e.g., Fig. A19), the oldest sandstone (and conglomerate) unit to house putative signs of life comes from the base of the ~3,400 Ma Strelley Pool Formation from the Pilbara of Western Australia (Fig. A24; Wacey et al., 2006). The geological context of this sandstone unit implies conditions conducive for life, with low angle cross bedding and channels. These, together with evidence for relatively

**Fig. A24** A typical early Archean sandstone from the base of the ~3,400 Ma Strelley Pool Formation, Pilbara, Western Australia (Photograph by the author)

high textural and compositional maturity indicate that deposition took place in a shallow beach-like setting, arguably the oldest such deposit in the rock record. This sandstone contains putative microborings into heavy mineral grains (Fig. B43), biologically mediated ambient inclusion trails (Fig. B45), cells and sheaths, possible biominerals in the form of micron-sized pyrite grains (Fig. B48), and possible ooids (Fig. B51). The evidence for life in this unit is critically reviewed on pages 181–190 and 195–196.

A slightly younger sandstone from the lower part of the ~3,200 – ~3,100 Ma Moodies Group in the Barberton greenstone belt, South Africa also contains evidence for early microbial life (Noffke et al., 2006). This sandstone is fine-grained and quartz-rich, has largely escaped deformation, and has only experienced low-grade metamorphism. The sandstone is interpreted as having been deposited across some of Earth's oldest tidal flats and the siliciclastic sediment appears to have been stabilized by microbial mats, now evidenced by wrinkle and roll up structures (Fig. B76). The evidence for life in this unit is critically reviewed on pages 229–233.

### *4.3.4 Hydrothermal Deposits*

Although most secondary Archean cherts appear to have a hydrothermal component (see Section 4.3.1), 'hydrothermal deposits' are here distinguished as those closely associated with hydrothermal vents. Hydrothermal vents (Fig. A25) that produce chimneys of (microbially?) precipitated iron sulphide are now a well known phenomenon of the modern deep sea floor along mid oceanic ridges (known as 'black

**Fig. A25** A type of hydrothermal vent known as a 'black smoker'. The black colour is due to the large amount of iron sulphide contained in the superheated soup of chemicals being ejected into the ocean from these cracks in the ocean floor. The chimney shape is formed by precipitation of (predominantly) iron sulphide minerals on contact with the cold seawater; partial oxidation of these causes the observed 'rust' colour. 'White smokers' are also common in hydrothermal vent fields, their colour being due to barium and calcium minerals (Image courtesy of NOAA)

smokers', e.g., Corliss et al., 1979) and back arc tectonic settings (e.g., Fouquet et al., 1991). While sulphide-rich black cherts are well known from hydrothermal rocks some 3,500–3,400 Ma old in Australia (e.g., Brasier et al., 2005), it is not until much later, in drill cores through the ~3,240 Ma Sulphur Springs Group of the Pilbara (Vearncombe et al., 1995) that hydrothermal black smoker deposits are convincingly preserved in the rock record. The Sulphur Springs sulphide deposit is associated with a sequence of komatiites, basalts, dacites and rhyolites that erupted on the seafloor (Van Kranendonk, 2006). Thin sections through the well-preserved drill core materials from the Sulphur Springs region show a wide range of characteristic volcanic and hydrothermal fabrics, including laminated pyrite nodules, dendritic and colloform sulphides, chalcedonic silica and vein quartz (e.g., Vearncombe et al., 1995).

Importantly, the Sulphur Springs deposit also provides putative evidence for life in the form of pyritic filaments from a massive sulphide unit which have been interpreted as the fossilised remains of thread like thermophilic, chemotrophic prokaryotes (Fig. B71; Rasmussen, 2000). A second piece of evidence comes from a diamond drill core passing through the Sulphur Springs group sediments. Here, bedding parallel filamentous organic bundles occur with bulk $\delta^{13}C$ values of between −26.8 and −34‰ (Fig. B72; Duck et al., 2007). It is not clear as to exactly which unit the organic bundles come from but they appear to be almost directly above the massive sulphide deposit described by Rasmussen (2000). The tubular filamentous bundles have been compared with those found around the Mid Atlantic Ridge today where the hyperthermophilic archeaon *Pyrodictium abysii* creates mats (Fig. B74; Rieger et al., 1995).

Evidence for life has also been reported from a hydrothermal setting in rocks of a similar age from the West Pilbara Superterrane. Here, the Dixon Island Formation of the 'Cleaverville Group' has been interpreted as a low temperature hydrothermal vent deposit (Kiyokawa et al., 2006). Black cherts within this Formation contain carbonaceous matter, in the form of filaments (Figs. B79, B80 and B82), spheres (Fig. B81) and mat-like laminations (Fig. B79). These have been interpreted as remains of a thermophilic, anoxygenic microbial community (Kiyokawa et al., 2006).

## Recommended Reading for Archean Rock Types


Brasier, M. D., Green, O. R., and Mcloughlin, N., 2004, Characterization and critical testing of potential microfossils from the early Earth: the Apex 'microfossil debate' and its lessons for Mars sample return, *International Journal of Astrobiology* **3**: 1–12.

Brasier, M. D., Green, O. R., Jephcoat, A. P., Kleppe, A. K., Van Kranendonk, M. J., Lindsay, J. F., Steele, A., and Grassineau, N. V., 2002, Questioning the evidence for Earth's oldest fossils, *Nature* **416**: 76–81.

Brasier, M. D., Green, O. R., Lindsay, J. F., McLoughlin, N., Steele, A., and Stoakes, C., 2005, Critical testing of Earth's oldest putative fossil assemblage from the ~3.5 Ga Apex Chert, Chinaman Creek Western Australia, *Precambrian Research* **140**: 55–102.

Brasier, M. D., McLoughlin, N., and Wacey, D., 2006, A fresh look at the fossil evidence for early Archaean cellular life, *Philosophical Transactions of the Royal Society B* **361**: 887–902.

Banerjee, N. R., Simonetti, A., Furnes, H., Muehlenbachs, K., Staudigel, H., Heaman, L., and Van Kranendonk, M. J., 2007, Direct dating of microbial ichnofossils, *Geology* **35**: 487–490.

Corliss, J. B., Dymond, J., Gordon, L. I., Edmond, J. M., Herzen, R. P. V., Ballard, R. D., Green, K., Williams, D., Bainbridge, A., Crane, K., and Van Andel, T. H., 1979, Submarine thermal springs on the Galapagos Rift, *Science* **203**: 1073–1083.

Duck, L. J., Glikson, M., Golding, S. D., and Webb, R. E., 2007, Microbial remains and other carbonaceous forms from the 3.24 Ga Sulphur Springs black smoker deposit, Western Australia, *Precambrian Research* **154**: 205–220.

Folk, R. L., and Weaver, C. E., 1952, A study of the texture and composition of chert, *American Journal of Science* **250**: 498–510.

Fouquet, Y., Vonstackelberg, U., Charlou, J. L., Donval, J. P., Foucher, J. P., Muhe, R., Wiedicke, M., Soakai, S., and Whitechurch, H., 1991, Hydrothermal activity in the Lau Back-Arc Basin – Sulfides and water chemistry, *Geology* **19**: 303–306.

Furnes, H., Banerjee, N. R., Muehlenbachs, K., Staudigel, H., and de Wit, M., 2004, Early life recorded in Archean pillow lavas, *Science* **304**: 578–581.

Garcia-Ruiz, J. M., Hyde, S. T., Carnerup, A. M., Christy, A. G., Van Kranendonk, M. J., and Welham, N. J., 2003, Self-assembled silicacarbonate structures and detection of ancient microfossils, *Science* **302**: 1194–1197.

Kiyokawa, S., Ito, T., Ikehara, M., and Kitajima, F., 2006, Middle Archean volcano-hydrothermal sequence: Bacterial microfossil-like bearing 3.2 Ga Dixon Island Formation, coastal Pilbara terrane, Australia, *Bulletin of the Geological Society of America* **118**: 3–22.

Lowe, D. R., 1999, Petrology and sedimentology of cherts and related silicified sedimentary rocks in the Swaziland Supergroup, In: Lowe, D. R., and Byerly, G. R. (eds.), *Geologic evolution of the Barberton Greenstone Belt, South Africa, Geological Society of America Special Paper* **329**: 83–114.

Noffke, N., Eriksson, K. A., Hazen, R. M., and Simpson, E. L., 2006, A new window into Early Archean life: microbial mats in Earth's oldest siliciclastic tidal deposits (3.2 Ga Moodies Group, South Africa), *Geology* **34**: 253–256.

Rasmussen, B., 2000, Filamentous microfossils in a 3235-million-year- old volcanogenic massive sulphide deposit, *Nature* **405**: 676–679.

Rieger, G., Rachel, R., Hermann, R., and Stetter, K. O., 1995, Ultrastructure of the hyperthermophilic Archaeon *Pyrodictium abyssi*, *Journal of Structural Biology* **115**: 78–87.

Schopf, J. W., 1993, Microfossils of the Early Archaean Apex Chert: new evidence for the antiquity of life, *Science* **260**: 640–646.

Siever, R., 1992, The silica cycle in the Precambrian, *Geochimica et Cosmochimica Acta* **56**: 3265–3272.

Sugitani, K., 1992, Geochemical characteristics of Archean cherts and other sedimentary rocks in the Pilbara Block, Western Australia: evidence for Archean seawater enriched in hydrothermally-derived iron and silica, *Precambrian Research* **57**: 21–47.

Staudigel, H., Furnes, H., Banerjee, N. R., Dilek, Y., and Muehlenbachs, K., 2006, Microbes and volcanoes: A tale from the oceans, ophiolites and greenstone belts, *GSA Today* **16**: 4–11.

Tice, M. M., and Lowe, D. R., 2004, Photosynthetic microbial mats in the 3,416-Myr-old ocean, *Nature* **431**: 549–552.

Vearncombe, S., Barley, M. E., Groves, D. I., McNaughton, N. J., Mikucki, E. J., and Vearncombe, J. R., 1995, 3.26 Ga black smoker type mineralization in the Strelley Belt, Pilbara Craton, Western Australia, *Journal of the Geological Society of London* **152**: 587–590.

Van Kranendonk, M. J., 2006, Volcanic degassing, hydrothermal circulation and the flourishing of early life on Earth: new evidence from the Warrawoona Group, Pilbara Craton, Western Australia, *Earth-Science Reviews* **74**: 197–124.

Van Kranendonk, M. J., and Pirajno, F., 2004, Geological setting and geochemistry of metabasalts and alteration zones associated with hydrothermal chert-barite deposits in the ca. 3.45 Ga Warrawoona Group. Pilbara Craton, Australia, *Geochemistry: Exploration, Environment, Analysis* **4**: 253–278.

Van Kranendonk, M. J., Webb, G. E., and Kamber, B. S., 2003, Geological and trace element evidence for a marine sedimentary environment of deposition and biogenicity of 3.45 Ga stromatolitic carbonates in the Pilbara Craton, and support for a reducing Archean ocean, *Geobiology* **1**: 91–108.

Wacey, D., McLoughlin, N., Green, O. R., Parnell, J., Stoakes, C. A., and Brasier, M. D., 2006, The ~3.4 billion-year-old Strelley Pool sandstone: a new window into early life on Earth, *International Journal of Astrobiology* **5**: 333–342.

Westall, F., de Vries, S. T., Nijman, W., Rouchon, V., Orberger, B., Pearson, V., Watson, J., Verchovsky, A., Wright, I., Rouzaud, J. N., Marchesini, D., and Severine, A., 2006, The 3.446 Ga "Kitty's Gap Chert", an early Archean microbial ecosystem, *GSA Special Paper* **405**: 105–131.

## Recommended Reading for Pilbara Geology

Buick, R., Thornett, J. R., McNaughton, N. J., Smith, J. B., Barley, M. E., and Savage, M., 1995, Record of emergent continental crust ~3.5 billion years ago in the Pilbara craton of Australia, *Nature* **375**: 574–577.

Hickman, A. H., 1983, Geology of the Pilbara block and its environs, *Western Australia Geological Survey Bulletin* **127**: 268.

Hickman, A. H., 2004, Two contrasting granite-greenstone terranes in the Pilbara Craton, Australia: evidence for vertical *and* horizontal tectonic regimes prior to 2900 Ma, *Precambrian Research* **131**: 153–172.

Kiyokawa, S., and Taira, A., 1998, The Cleaverville Group in the West Pilbara coastal granitoid-greenstone terrain of Western Australia: an example of a Middle Archean immature oceanic island-arc succession, *Precambrian Research* **88**: 109–142.

Lowe, D. R., 1983, Restricted shallow-water sedimentation of early Archaean stromatolitic and evaporitic strata of the Strelley Pool chert, Pilbara block, Western Australia, *Precambrian Research* **19**: 239–283.

Smithies, R. H., Van Kranendonk, M. J., and Champion, D. C., 2005, It started with a plume – early Archaean basaltic proto-continental crust, *Earth and Planetary Science Letters* **238**: 284–297.

Van Kranendonk, M. J., 2006, Volcanic degassing, hydrothermal circulation and the flourishing of early life on Earth: new evidence from the Warrawoona Group, Pilbara Craton, Western Australia, *Earth-Science Reviews* **74**: 197–124.

Van Kranendonk, M. J., Hickman, A. H., Williams, I. R., and Nijman, W., 2001, Archean Geology of the East Pilbara Granite-Greenstone Terrane Western Australia– A Field Guide, *Western Australia Geological Survey Record* **2001/9**: 134 p.

Van Kranendonk, M. J., Hickman, A. H., Smithies, R. H., and Nelson, D., 2002, Geology and tectonic evolution of the Archean North Pilbara Terrain, Pilbara Craton, Western Australia, *Economic Geology* **97**: 695–732.

Van Kranendonk, M. J., Hickman, A. H., and Huston, D. L., 2006, Geology and mineralization of the East Pilbara – a field guide, *Western Australia Geological Survey Record* **2006/16**: 90 p.

Van Kranendonk, M. J., Smithies, R. H., Hickman, A. H., and Champion, D. C., 2007, Review: secular tectonic evolution of Archean continental crust: interplay between horizontal and vertical processes in the formation of the Pilbara Craton, Australia, *Terra Nova* **19**: 1–38.

Wacey, D., McLoughlin, N., Green, O. R., Parnell, J., Stoakes, C. A., and Brasier, M. D., 2006, The ~3.4 billion-year-old Strelley Pool sandstone: a new window into early life on Earth, *International Journal of Astrobiology* **5**: .333–342.

Wacey, D., McLoughlin, N., Stoakes, C. A., Kilburn, M. R., Green, O. R., and Brasier, M. D., 2008, The ~3.4 Ga Strelley Pool Chert in the East Strelley greenstone belt– a field and petrographic guide, *Western Australia Geological Survey Record.*

# Recommended Reading for Barberton Geology

Anhaeusser, C. R., 1973, The evolution of the early Precambrian crust of South Africa, *Philosophical Transactions of the Royal Society of London A* **273**: 359–388.

Armstrong, R. A., Compston, W., de Wit, M. J., and Williams, L. S., 1990, The stratigraphy of the 3.5–3.2 Ga Barberton Greenstone Belt revisited: a single zircon ion microprobe study, *Earth and Planetary Science Letters* **101**: 90–106.

Hessler, A. M., and Lowe, D. R., 2006, Weathering and sediment generation in the Archean: an integrated study of the evolution of siliciclastic sedimentary rocks of the 3.2 Ga Moodies Group, Barberton Greenstone Belt, South Africa, *Precambrian Research* **151**: 185–210.

Heubeck, C., and Lowe, D. R., 1994, Depositional and tectonic setting of the Archean Moodies Group, Barberton Greenstone Belt, South Africa, *Precambrian Research* **68**: 257–290.

Hofmann, A., 2005, The geochemistry of sedimentary rocks from the Fig Tree Group, Barberton greenstone belt: implications for tectonic, hydrothermal and surface processes during mid-Archaean times, *Precambrian Research* **143**: 23–49.

Hofmann, A., and Bolhar, R., 2007, Carbonaceous cherts in the Barberton Greenstone Belt and their significance for the study of early life in the Archean record, *Astrobiology* **7**: 355–388.

Knauth, L. P., and Lowe, D. R., 2003, High Archean climatic temperatures inferred from oxygen isotope geochemistry of cherts in the 3.5 Ga Swaziland Supergroup, South Africa, *Bulletin of the Geological Society of America* **115**: 566–580.

Kroner, A., Hegner, E., Wendt, J. I., and Byerly, G. R., 1996, The oldest part of the Barberton granitoid-greenstone terrain, South Africa: evidence for crust formation between 3.5 and 3.7 Ga, *Precambrian Research* **78**: 105–124.

Lowe, D. R., and Byerly, G. R., 1999, Geologic Evolution of the Barberton Greenstone Belt, South Africa, *GSA Special Paper* **329**, Boulder, CO.

Zegers, T. E., de Wit, M. J., Dann, J., and White, S. H., 1998, Earth's oldest assembled continent? A combined structural, geochronological and palaeomagnetic test, *Terra Nova* **10**: 250–259.

# Recommended Reading for South-West Greenland Geology

Appel, P. W. U., Moorbath, S., and Touret, J. L. R., 2003, Early Archaean processes and the Isua Greenstone Belt, West Greenland, *Precambrian Research* **126**: 173–179.

Dauphas, N., van Zuilen, M., Wadhwa, M., Davis, A. M., Marty, B., and Janney, P. E., 2004, Clues from Fe isotope variations on the origin of early Archean BIFs from Greenland, *Science* **306**: 2077–2080.

Fedo, C. M., and Whitehouse, M. J., 2002, Metasomatic origin of quartz-pyroxene rock, Akilia, Greenland, and its implications for Earth's earliest life, *Science* **296**: 1448–1452.

Kamber, B. S., Moorbath, S., and Whitehouse, M. J., 2001, The oldest rocks on Earth: time constraints and geological controversies, *Special Publication of the Geological Society of London* **290**: 177–203.

Lepland, A., Arrhenius, G., and Cornell, D., 2002, Apatite in the Early Archean Isua supracrustal rocks, southern West Greenland: its origin, association with graphite and potential as a biomarker, *Precambrian Research* **118**: 221–241.

Lepland, A., van Zuilen, M. A., Arrehnius, G., Whitehouse, M. J., and Fedo, C. M., 2005, Questioning the evidence for earth's earliest life – Akilia revisited, *Geology* **33**: 77–79.

Manning, C. E., Mojzsis, S. J., and Harrison, M. T., 2006, Geology, age and origin of supracrustal rocks at Akilia, west Greenland, *American Journal of Science* **306**: 303–366.

Mojzsis, S. J., Arrenhius, G., McKeegan, K. D., Harrison, T. M., Nutman, A. P., and Friend, C. R. L., 1996, Evidence for life on Earth 3,800 million years ago, *Nature* **384**: 55–59.

Moorbath, S., O'Nions, R. K., and Pankhurst, R. J., 1973, Early Archaean age for the Isua iron formation, West Greenland, *Nature* **245**: 138–139.

Nutman, A. P., McGregor, V. R., Friend, C. R. L., Bennett, V. C., and Kinny, P. D., 1996, The Itsaq Gneiss Complex of southern West Greenland; the world's most extensive record of early crustal evolution (3900–3600 Ma), *Precambrian Research* **78**: 1–39.

Nutman, A. P., Mojzsis, S. J., and Friend, C. R. L., 1997, Recognition of ≥3850 Ma water-lain sediments in West Greenland and their significance for the early Archaean Earth, *Geochimica et Cosmochimica Acta* **61**: 2475–2484.

Rosing, M. T., 1999, $^{13}C$ Depleted carbon microparticles in >3700-Ma sea-floor sedimentary rocks from West Greenland, *Science* **283**: 674–676.

van Zuilen, M. A., Lepland, A., and Arhenius, G., 2002, Reassessing the evidence for the earliest traces of life, *Nature* **418**: 627–630.

van Zuilen, M. A., Lepland, A., Teranes, J., Finarelli, J., Wahlen, M., and Arrhenius, G., 2003, Graphite and carbonates in the 3.8 Ga old Isua Supracrustal Belt, southern West Greenland, *Precambrian Research* **126**: 331–348.

Whitehouse, M. J., and Fedo, C. M., 2007, Searching for Earth's earliest life in southern West Greenland – history, current status and future prospects. In: Van Kranendonk, M. J., Smithies, H., and Bennett, V. (Eds.) *Earth's Oldest Rocks. Developments in Precambrian Geology* **15**: 841–853.

# Chapter 5
# Techniques for Investigating Early Life on Earth

## 5.1 Introduction

It is generally agreed that three independent and mutually supporting lines of evidence are required for any claim of early life to be compelling:

- Evidence for a well-constrained age and context. This includes mapping at scales from kilometres to metres, supported by mapping of petrographic thin sections in order to show that candidate structures are truly syngenetic and ancient. Contextural and petrographic mapping is also necessary to falsify the 'null hypothesis' of a non-biological origin;
- Evidence for a morphology unique to biology. This includes *in situ* imaging and mapping of morphospace, to distinguish the fields of biological and non-biological morphology and to compare these with self-organising structures;
- Geochemical evidence for metabolic cycling. Geochemical evidence for life requires high-resolution 3D µm-scale to nano-scale *in situ* mapping and analysis, using more than a single line of contaminant-free investigation.

In the following section I highlight the major techniques presently used for the investigation of early life on Earth.

## 5.2 Geological Mapping

The geological context of samples and their spatial relationships are paramount in the study of early life. The context for potential biosignals should be mapped and studied at a wide range of scales from kilometres to micrometers. Evidence for, and interpretation of, the context should be clearly separated. Any potential biological signals should be referable to a well-defined history within this context, to show if they are early or late, indigenous or exogenous. Plausibility of the context for early life can then be assessed.

The importance of geological mapping is perhaps best demonstrated with the use of an example from the Pilbara of Western Australia. For many years the world

D. Wacey, *Early Life on Earth: A Practical Guide*,

famous 'microfossils' from the ~3,460 Ma Apex chert held a key place in Archean Palaeobiology, with 11 putative species of microfossils providing the oldest accepted morphological evidence for life on Earth. A detailed geological map was never presented to accompany this claim for life, and a poor understanding of the geological context proved to be a vital flaw in this work. It had been claimed that the microfossil locality was part of a bedded sequence of siliceous cherts (Schopf, 1993) comparable in part with modern day evaporitic and lagoonal environments. This interpretation, combined with the morphology of the putative microfossils, lead some to postulate that these earliest microfossils may be cyanobacteria, possibly even oxygen releasing cyanobacteria. If accepted, this would imply high levels of diversity soon after the end of heavy meteoritic bombardment, with little further diversification until the widespread emergence of eukaryotes nearly 2 billion years later (e.g., Knoll, 2003), plus an early start for the contribution of photosynthetic oxygen to the atmosphere.

These arguments sat uneasily with a group led by Martin Brasier from Oxford University. Their detailed mapping (Brasier et al., 2005, 2008) instead shows the 'microfossil site' to be 100 m down a large hydrothermal vein rather than in an equitable surface environment. While this does not entirely rule out the presence of life (hyper-thermophiles may well survive such conditions), it does preclude the presence of photosynthetic cyanobacteria and all that this implies.

## 5.3 Radiometric Dating

Radiometric dating is used to give an absolute age for a variety of different rock types. The technique relies upon the radioactive decay of unstable isotopes of particular elements (referred to as nuclides). The abundance of a naturally occurring nuclide (the 'parent') compared to its decay product (the 'daughter') is measured using a mass spectrometer. This ratio is then substituted into the 'age equation' together with the experimentally derived decay rate constant between parent and daughter nuclides, so that the measured abundances can be translated into ages:

$$t = 1/\lambda \ln (1 + D/P) \tag{5.1}$$

where

$t$ = age of the sample
$D$ = number of atoms of the daughter nuclide in the sample
$P$ = number of atoms of the parent nuclide in the sample
$\lambda$ = decay constant of the parent isotope (inversely proportional to the 'half life')

Many different isotopic systems are available for radiometric dating, and different systems vary in the timescale over which they are accurate and the materials to which they can be applied.

**Timescale**: Here the choice of isotopic system depends on the half life of the decaying nuclide. A given radioactive nuclide decays exponentially at a rate described by a parameter known as the 'half-life'. The half life is usually quoted in units of years and is essentially a constant for a given nuclide. After one half-life has elapsed, one half of the atoms of the nuclide in question will have decayed into a daughter nuclide. In many cases, the daughter nuclide itself is radioactive, resulting in a decay chain, eventually ending with the formation of a non-radioactive final daughter product; each step in the chain is characterized by a distinct half-life. For such chains, the half-life of interest in radiometric dating is usually the longest one in the chain, since this is the rate-limiting step in the ultimate transformation of the radioactive nuclide into its stable daughter.

Clearly, the precision of radiometric dating depends on choosing a system with a suitable half life, which in turn is dependent on the approximate age of the object that needs to be dated. The parent nuclide must have a long enough half-life that it will still be present in significant amounts at the time of measurement. Likewise, enough of the daughter product must be produced to be accurately measured and to be distinguished from any initial amount of daughter present in the material. For example, to date an archaeological find such as human remains, one would use carbon-14 dating with a $^{14}C$ half life of 5,730 years. In contrast, to date an Archean rock, $^{14}C$ would be useless because all of the parent nuclide would have decayed long before the present day. In this case, one would use either uranium–lead (U–Pb) dating ($^{235}U$ half life is 704 million years; $^{238}U$ half life is 4,470 million years) or perhaps potassium–argon (K–Ar; $^{40}K$ half life is ~1,300 million years) or rubidium–strontium (Rb–Sr; $^{87}Rb$ half life is 50 billion years) dating. Argon–argon ($^{40}Ar$–$^{39}Ar$) is also used as a proxy for K–Ar.

**Materials**: Here the choice of isotopic system depends on the rock type in question. The best minerals for radiometric dating are those that selectively incorporate a parent nuclide and reject the daughter nuclide during their formation (so that one can assume that all of the daughter nuclide measured is from radioactive decay of the parent), have a high 'closure temperature' (i.e., the temperature at which nuclides might enter or leave the crystal structure), and are stable to chemical and physical weathering. Zircon ($ZrSiO_4$) is probably the best example of such a mineral and is the mineral of choice for the U–Pb system. The U–Pb system is the most straightforward, gives the best precision (as good as ±2 Ma for rocks of 3,000 Ma), and also possesses a first order cross-check on data reliability because both the $^{235}U$ and $^{238}U$ clocks are operating simultaneously. In contrast, micas, feldspars and hornblendes can contain significant $^{40}$ K but their closure temperatures are much lower, meaning that the K–Ar 'clock' is often altered by metamorphism; remember that minerals usually only record the last time they cooled below their closure temperature.

When a mineral incorporates both parent and daughter nuclides during its formation a correction must be made for this initial ratio using an isochron plot. This is the case, for example, for rhenium–osmium (Re–Os, half life 41.6 billion years) dating of sulphides and for rubidium–strontium (Rb–Sr) dating of granitic rocks. In granites, mineral associations (plagioclase, K-feldspar, hornblende, biotite and muscovite) record evolving initial Rb–Sr ratios, and provide a cross check on consistency. Here, the age of a sample is determined by analysing several minerals; the $^{87}Sr/^{86}Sr$ ratio for each sample is plotted against its $^{87}Rb/^{86}Sr$ ratio on an isochron graph. If these form a straight line then the samples are consistent, the age is likely reliable, and the slope of the line indicates the actual age. The precision is about ±30 Ma for 3,000 Ma rocks. This method is unlikely to give an accurate formation age for rocks that have suffered significant hydrothermal alteration because both Rb and Sr are relatively mobile elements, hence easily redistributed by hot hydrothermal fluids. Where possible, it is always advisable to date more than one mineral from a rock, or date the same mineral using different isotopic systems. Apatite, monazite, titanite, epidote and garnet are other useful dating minerals.

Radiometric dating is performed using a mass spectrometer which can accurately measure isotopic ratios on separated grains, or *in situ* on crystals as small as a few microns. The *in situ* technique, using SIMS (see Section 5.7), lasers or SHRIMP (sensitive high resolution ion microprobe) is particularly useful for minerals which are zoned (this is common in zircons) – each of these zonations may be dated and may record, for example, a series of metamorphic events. Radiometric dating has revolutionised many areas of geology, but its single most important contribution is the assignment of absolute ages to periods and events on the geological column (Fig. A26). Recent technological advancements have improved the precision of many of these dates to around ±2 million years, even for 3,000 million-year-old plus rocks.

An interesting and relatively new dating application uses extinct radioactive nuclides. Examples of these include $^{26}Al$, $^{60}Fe$, $^{53}Mn$, and $^{129}I$ which were only present in the earliest history of planet formation. Their decay products are occasionally found in extremely old material, the best example being meteorites, and this enables the determination of the *relative* ages between different events in the early history of the solar system. This relative history can then be calibrated to some degree using U–Pb dating to give *absolute* ages. Samarium–neodymium ($^{147}Sm$–$^{143}Nd$, half life 106 billion years) dating is also useful for determining the age relationships of rocks and meteorites during the earliest history of our planet and has been applied particularly to investigating crustal formation ages.

## 5.4 Optical Microscopy

The analysis of putative biological structures in a petrographic thin section is a vital part in the investigation of early life. Optical microscopy (also referred to as petrography) can be used to reveal the morphology of the structure of interest in two and sometimes three dimensions. Petrography acts as an essential extension of geological mapping because it allows the context of any structure to be mapped on the millimetre to micron scale. Various types of equipment can be used to enhance a standard optical microscope to obtain the maximum information from a given sample using plane polarised transmitted light, cross polarised transmitted light, reflected light, cathodoluminescence, lasers and various digital imaging and processing packages.

One technique that has vastly improved the amount of information one can convey in images is *Auto-montage*™. This is a sophisticated image acquisition and processing software package, and is an ideal tool in the examination of a variety of microstructures

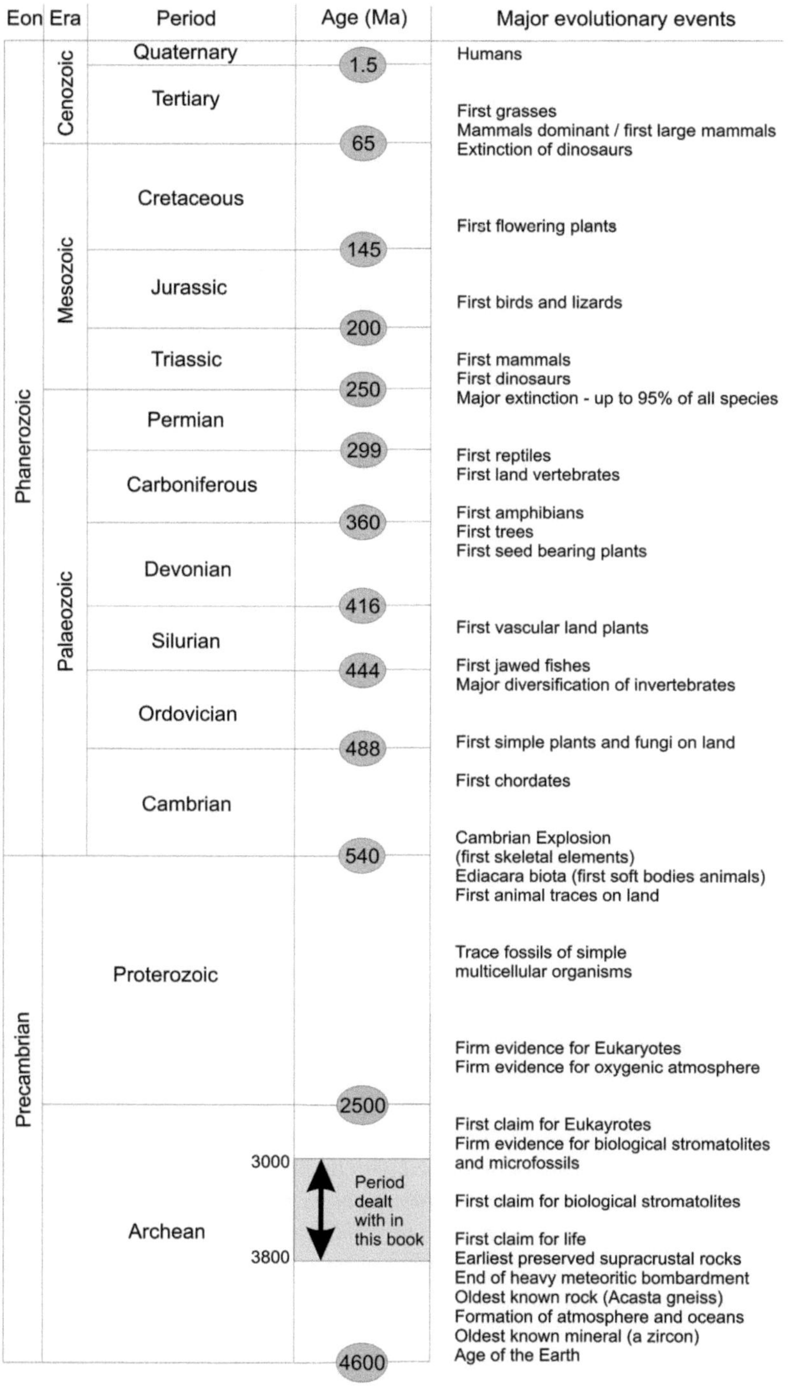


**Fig. A26** The geologic column, highlighting the usefulness of radiometric dating in order to place absolute age values on the important evolutionary events that have shaped our planet within standard and non-standard thin sections. With it, the user is able to capture, process and combine multiple source images, each obtained at different focal depths within a thin section. Processing algorithms enable the most sharply focussed areas of each source image to be combined into a single well-focused 'montaged' image. This rendering facility is ideal for obtaining high-resolution images of 3-D microscopic structures aligned obliquely to the z-axis within a thin section. It is similar to the established technique of manual montage, in which images of a microstructure are collected by optical photomicrography and then manually spliced in the darkroom or laboratory. It differs, however, in that entire selected focal planes from the 'montaged' image can also be displayed on the screen or print. Depth maps, confidence maps and 3-D images can also be generated from the series of 2-D source images. Figures A27 and A28 demonstrate the use of *Auto-montage*™ using standard thin sections containing putative microfossils from the ~3,460 Ma Apex chert and the ~3,400 Ma Strelley Pool Formation of Western Australia. It can be seen that when images from successive focal planes are combined, a sharp and accurate montage is obtained.

## 5.5 Scanning Electron Microscopy (SEM)

SEM provides topographic and compositional images of the surface of a sample. The scanning electron microscope (also known as a SEM) can accommodate numerous types of sample making it a highly versatile machine. Samples can include thin sections, crushed or powdered rock samples mounted onto stubs, individual or multiple mineral crystals, and larger (several millimeters) rock chips. A sample surface is imaged by scanning it with a high-energy beam of electrons in a raster scan pattern. When the electrons hit the sample surface they interact with the atoms of the sample producing numerous secondary signals. These include **secondary electrons**, **back scattered electrons**, **x-rays** and **cathodoluminescence**.

**Secondary electrons** (SE) are low energy electrons (<50 eV) ejected from the sample atoms at a very shallow depth (top few nanometers) within the sample. These give very high resolution images of the sample surface topography. The spatial resolution can be as good as 1–10 nm. These images have a very large depth of focus, preserving any three-dimensional features of a sample (although to obtain a true three-dimensional image the user must combine several images taken from different angles). For early life studies, siliceous samples are often etched with hydrofluoric acid (HF) prior to examination to enhance the relief of the sample; carbonaceous microbial structures would then be seen to stand proud from the surrounding matrix (e.g., Fig. B81) because HF fumes preferentially attack pure $SiO_2$. The SEM also allows a wide range of magnifications to be chosen, typically around five orders of magnitude from about ×25 to ×250,000. This variation allows both coarse scale contextural mapping and fine scale morphological imaging to be carried out in quick succession on the same sample.

**Back scattered electrons** (BSE) are emitted from atoms slightly deeper within the sample and are of higher energy. The yield of BSE from a given atom is proportional to the atomic number and density of that excited atom. BSE thus provide qualitative

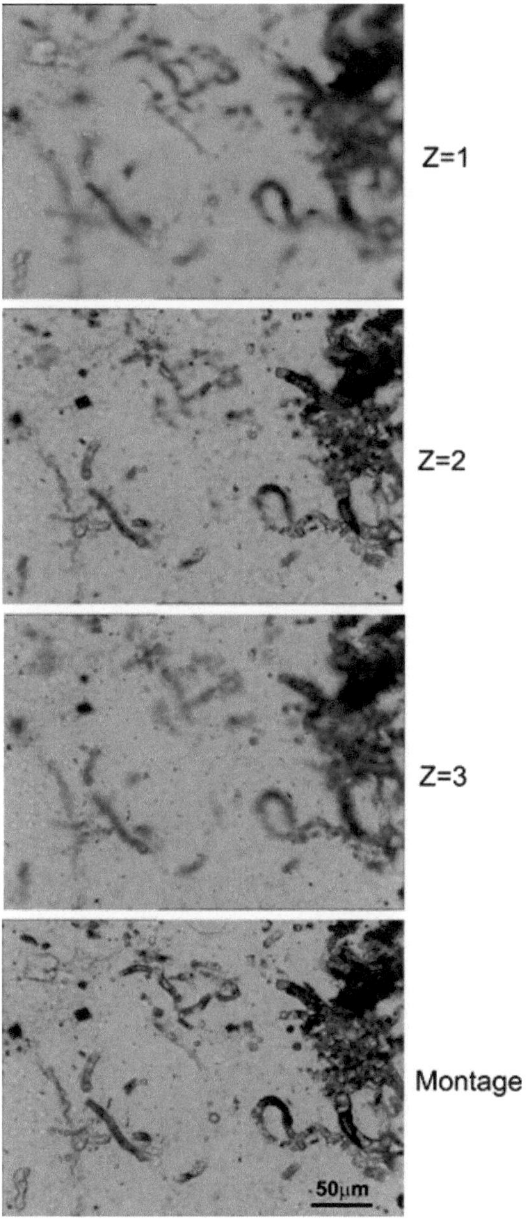


**Fig. A27** A demonstration of the *Auto-Montage*™ technique. Images are taken from numerous z-axis focal planes and combined into one montaged image. Sample is from the ~3,400 Ma Strelley Pool sandstone held in the Oxford Palaeobiology Collections

information on the chemical composition of the sample. The brightness of a BSE image tends to increase with increasing atomic number and these images are usually used in combination with x-rays (below) to work out chemical composition. BSE can also be used to form an electron backscatter diffraction image which can be used to determine the crystallographic structure of the sample.

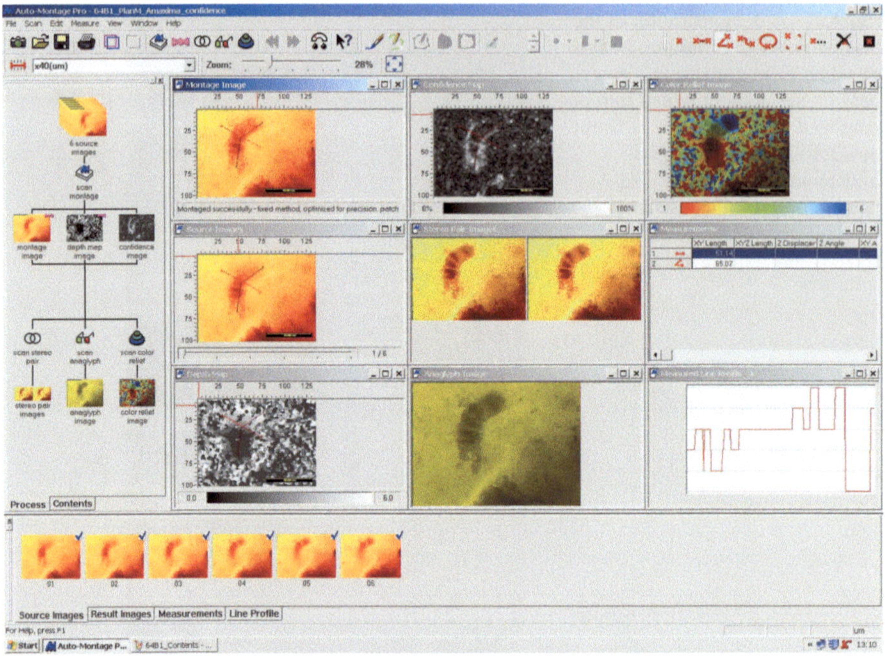


**Fig. A28** Screen captures from the *Auto-Montage*™ system. Six source images from different focal depths are seen in the bottom panel. The remaining panels show various manipulations possible using the montaged image. These include depth profiles, confidence intervals, stereo pairs and linear measurements (Sample is from the ~3,460 Ma Apex chert courtesy of the Natural History Museum, London)

**X-rays** are emitted from a sample when the incident electron beam removes an inner shell electron from a sample atom, causing a higher energy electron to fill the shell and give off its additional energy. X-rays are produced at depths of up to 1 μm or more (dependent on the energy of the electron beam) within the sample. These x-rays produce characteristic spectra indicating which chemical elements are present in the analysed area (this technique is generally known as **energy dispersive x-ray spectrometry** (EDXS) ).

**Cathodoluminescence** (CL). Certain materials (notably for studies of early life those containing phosphorous) release excess energy in the form of photons of light when electrons recombine to fill holes made by the excitation of the sample by the primary beam. In the SEM, CL detectors collect and analyze the wavelengths emitted by the specimen and display an emission spectrum (which can be compared to known standard mineral spectra) or an image of the distribution of CL within the specimen. Image resolution using CL can be as good as about 50 nm.

There are, however, some significant limitations to the SEM for studies of early life. Low atomic mass elements (including carbon and nitrogen) are very difficult to quantify using EDXS. This is because the lower the atomic mass of an atom is, the lower its x-ray yield is upon excitation. There is also a large discrepancy between the spatial resolution achievable for imaging versus chemical analysis. For example, secondary electrons can give nano-scale information about the upper few nanometers of a sample, whereas x-rays give information only on a 1 μm or so scale. It is almost impossible, therefore, to obtain chemical compositions of nanometre-scale objects.

Early palaeontological studies using the SEM were plagued by contamination and artefact problems (e.g., Brooks et al., 1973). In recent years, however, improvements in SEM technology combined with a better understanding of the way bacteria can bc fossilized has led to its re-introduction into early life studies (e.g., Westall et al., 2001; Westall and Folk, 2003). The SEM has recently provided *in situ*, high resolution images of putative microbial structures (e.g., Fig. A29). Most SEM can rotate and tilt a sample, allowing a 3 dimensional picture of the microbial structure and its relationship to the surrounding rock matrix to be built up. In this way SEM can be used to evaluate whether a given structure is indigenous to the rock or contained within cracks or pores (e.g., B95 top). Nevertheless, great care must be taken not to introduce contamination during sample preparation (including acid etching), and it is imperative that structures seen under the SEM are also observed in standard petrographic thin sections.

SEM sample preparation is relatively straight forward. For good BSE imaging and quantitative x-ray analysis/mapping the sample surface must be ground and polished to an ultra smooth surface. Insulating specimens (i.e., most rocks and biological materials) must be coated with a conductive material to prevent the localized build of electron charge. 'Charging' degrades the image and can even lead to artefacts in elemental spectra. The conductive coat commonly consists of a thin layer (a few tens of nm) of carbon or gold, but platinum, palladium, tungsten and iridium are increasingly being used. For EDXS analysis carbon is the preferred coating material,

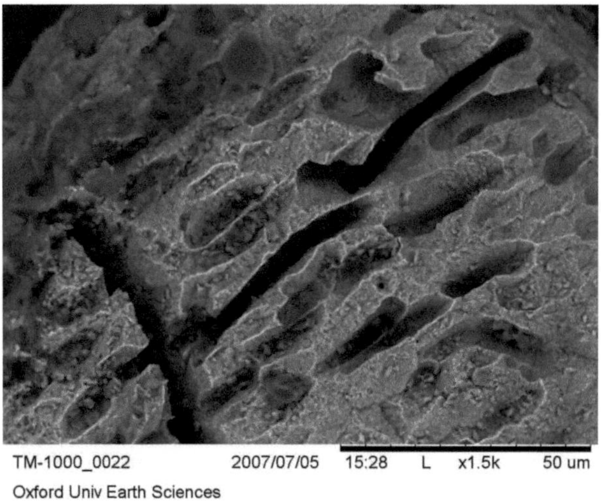


**Fig. A29** SEM image of micro-tubular structures within a detrital pyrite grain from the ~3,400 Ma Strelley Pool sandstone. Whilst the SEM provides a morphological argument for the biogenicity of these structures, further petrographic and geochemical evidence for biogenicity and antiquity would need to be obtained using other techniques (see *below*)

since the heavy metal coats may obscure some of the lower atomic number spectra that the user is trying to resolve. In addition to coating the sample, the specimen should be mounted on a stub in such a way that a good electrical path is established. This is usually accomplished through the use of a conductive adhesive such as silver or colloidal carbon paint.

For modern biological samples (e.g., cultured microbes, modern stromatolites) it may be necessary to 'fix' the sample to preserve its structure. This may be achieved chemically by incubation in gluteraldehyde or formaldehyde, by dehydration, or by freezing (cryofixation). Biological samples are also often embedded in resin prior to sectioning and polishing for the SEM.

Low vacuum or **environmental scanning electron microscopes** (ESEM) can be used to eliminate much of the need for coating. Here the specimen is placed in an internal chamber maintained at higher pressure than the vacuum in the electron optical column. Positively charged ions generated by beam interactions with the gas help to neutralize the negative charge on the specimen surface. The pressure of gas in the chamber can be controlled, and the type of gas used can be varied according to the need of the user. This is especially useful when dealing with delicate non-conductive biological material where coating may conceal important nano-scale features. It is also very useful when the user has limited amounts of a key sample, and coating would inhibit further analysis using other techniques. ESEM also makes it possible to perform EDXS chemical analyses on uncoated non-conductive specimens. The user must, however, sacrifice a certain amount of resolution when using an ESEM compared to a conventional SEM.

## 5.6 Transmission Electron Microscopy (TEM)

TEM can be used to image and characterize the crystallography and chemistry of a sample at the nanometre scale. In TEM, a beam of electrons is transmitted *through* a sample, interacting with the atoms as it passes through. There is thus the requirement for the sample to be ultra-thin (usually <100 nm for high resolution work). As the electron beam passes through the sample it is diffracted; the amount of the diffraction depends on the orientation of the planes of atoms within the sample in relation to the electron beam (determined by Bragg's law). In the transmission electron microscope (also known as a TEM), a sample can be tilted in order to achieve specific diffraction conditions, and apertures can be placed below the sample in order to collect electrons diffracted in specific directions.

Crystal structure is best investigated by high resolution transmission electron microscopy (HRTEM). This is also known as phase contrast imaging because the images are formed due to differences in the phases of electron waves diffracted through the specimen. The spatial resolution of HRTEM is phenomenal, producing images with sufficient resolution to show carbon atoms in diamond separated by only 0.89 Å (89 pm, 1 Å is 0.0000000001 of a meter or 100 pm) at magnifications of 50 million times. In the early life field, Rouzaud et al. (2005) have directly imaged the polyaromatic layers within carbon from materials ranging from meteorites to the ~3,460 Ma Apex chert. HRTEM was able to show the length and stacking arrangement of these layers on the nano-metre scale. Once a sufficient database of standard materials is built up, such length and stacking arrangements may be able to distinguish between biological and non-biological carbon.

Chemical analyses can be made by using an add-on scanning mode (scanning transmission electron microscopy, or STEM). In STEM, energy dispersive x-ray analyses (EDX) can be made in much the same way as with the SEM. However, the problem of the depth of x-ray production limiting the resolution in SEM is not encountered in STEM because the samples are so thin. Thus, STEM can produce chemical maps of a sample at nanometre scale resolution.

The major drawback to TEM work comes with the time consuming and destructive nature of the sample preparation. Target areas must first be selected in thin sections using optical microscopy and/or SEM. Next, cross sections through these targets must be extracted and thinned to ~100 nm. This is usually done by focussed ion beam (FIB) milling, using a highly focussed beam of gallium ions to effectively cut through the sample at the atomic scale. Because of the destructive nature of this preparation technique and the small field of view in the TEM, the user must be confident that the context of any putative microfossil has been thoroughly and accurately mapped before FIB milling – there is no going back to check!

## 5.7 Secondary Ion Mass Spectrometry (SIMS and NanoSIMS)

SIMS is used to determine the elemental, isotopic or molecular composition of a sample surface. The surface of a specimen is sputtered with a focused primary ion beam and ejected secondary ions are collected and analyzed using a mass spectrometer. SIMS is one of the most sensitive surface analysis techniques, being able to detect elements present in the parts per billion range.

The NanoSIMS is a new generation of dynamic SIMS which enables elements to be mapped at nanometer scale resolution (Fig. A30), and isotopic measurements to be made *in situ* from micron to sub-micron sized objects. A primary ion beam is scanned across the surface of the sample, and the sputtered secondary ions are extracted to a double focusing mass spectrometer. The nanoSIMS is capable of sub-50 nm lateral resolution whilst imaging negatively charged secondary ions (using $Cs^+$ primary ions) and sub-150 nm lateral resolution for positive secondary ions (using $O^-$ primary ions). Four moveable and one static detector record ion counts from the same sputtered sample volume at five masses simultaneously. This parallel measurement is essential in the analysis of low concentration elements in organic materials that are rapidly eroded by the ion beam.

Isotope measurements are made by tuning the mass spectrometer to high-mass -resolution. This is achieved through the use of slits at the entrance to the mass-spectrometer, and deflector plates directly in front of the detectors. With high mass resolving power the user can, for example, easily resolve $^{13}C$ from $^{12}C^{1}H$ whilst

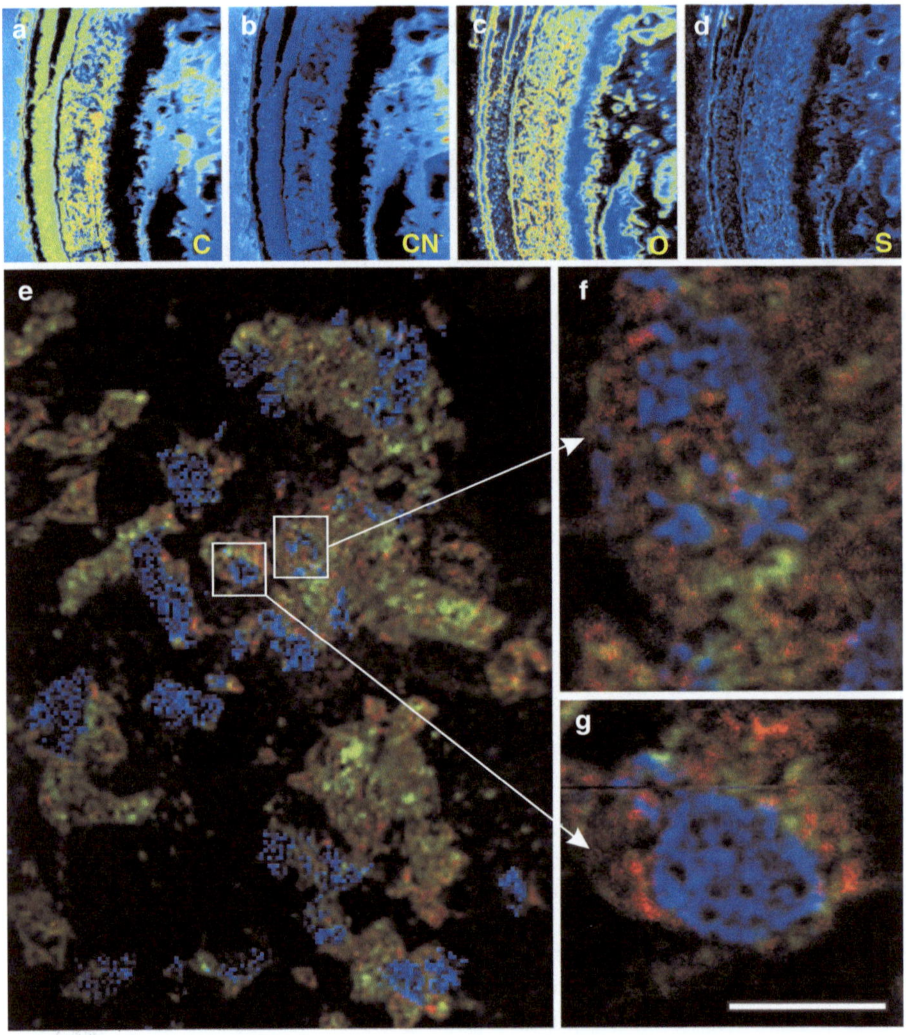


**Fig. A30** NanoSIMS images of modern and ancient biosignals. **a–d**, Low resolution NanoSIMS images of carbon, nitrogen (as $CN^-$), oxygen and sulphur, illustrating the distribution of these elements in part of a modern lithified microbial mat from Shark Bay, Western Australia. Relative elemental abundance is shown by intensity of colour from dark blue (lowest) through light blue, green, yellow to orange/red (highest). Areas where carbonate precipitation has occurred are shown by a positive correlation between carbon and oxygen maps. Areas where organic material is present are shown by anti-correlation of oxygen with carbon and nitrogen maps (e.g., far left and centre right of figure). **e-g**, High resolution NanoSIMS images of carbon (red), nitrogen (green) and pyrite (blue) showing clumps of organic material attached to micron-sized pyrite grains in the ~3400 Ma Strelley Pool sandstone. Samples are from the Oxford Palaebiology Collections. Imaging carried out at CMCA with the help of Matt Kilburn. Scale bar is 50 microns for a-d; 12 microns for e; and 2 microns for f-g maintaining a good number of counts. Measurements are made by rastering a slightly defocused $Cs^+$ primary beam over a given area. The signal from the central pixels is recorded, while the outer pixels are discarded to eliminate the possible contribution from the edges of the crater caused by the ion beam.

NanoSIMS is a new technique to the field of early life and much of the current work is focused upon achieving the required reproducibility and precision for isotopic work (especially carbon and sulphur) on putative biological material. Nevertheless some interesting data has already emerged with high resolution maps of carbon, nitrogen, and other biologically important elements being used to argue for biological participation in ~3,400 Ma ambient inclusion trails (Wacey et al., 2008). Both carbon and sulphur isotypes have been measured to a precision of about 3 per mil in these ~3,400 Ma samples, and this appears adequate to trace biological fractionations. NanoSIMS has also been used to investigate putative biological material from meteorites with preliminary results showing direct correlations between C and CN in the Nakhla meteorite (Gibson Jr. et al., 2006). NanoSIMS is also revealing valuable new information from younger microfossil suites. In the 850 Ma Bitter Springs chert, for example, filaments and spheres with distinct wall and sheath structures have been recognised in organic material that was previously thought to be amorphous (Oehler et al., 2006).

Limitations of the nanoSIMS include the inability to image or measure anything that is not at or very close to the surface of a thin section, and the limited precision of isotopic analysis (for example, one would be ill advised to attempt iron isotope measurements using nanoSIMS because of the small fractionations involved).

## 5.8 Laser-Raman Micro-Spectroscopy

Laser-Raman micro-spectroscopy allows the collection of molecular and structural data, in situ, at high sensitivity and micron-scale spatial resolution. It is non-intrusive and non-destructive, and can be used on bulk samples or standard uncovered thin sections, making it an ideal technique for samples containing rare putative microfossils. Analyses can be made from objects both at the surface and at depth within a thin section (although the best images are produced from objects at or just below the surface), meaning that direct chemical and optical comparisons can be made. Should further techniques be required, they can be carried out on the identical area as the laser-Raman (because of the non-destructive nature of the technique) so that extra information (e.g., isotopic composition) can be overlain on Raman images. Modern laser-Raman systems have a standard microscope body so that, with a simple flick of a switch, imaging in transmitted and reflected light at a variety of magnifications can be performed in parallel to Raman data acquisition. Raman data is acquired by laser excitation of the samples using wavelengths typically ranging between the blue and infrared parts of the spectrum. Data can be presented as both point spectra and 2-D or 3-D images.

**Point spectra**: Raman point spectra give information on molecular bonding and crystallinity in many minerals. One application of this is to quickly and simply identify different mineral phases within a sample. Of particular interest in studies of early life is the identification and characterisation of carbon-rich material associated with putative microfossils, trace fossils, laminae or merely ancient carbonaceous blobs.

The Raman spectra of various types of carbonaceous material have been widely studied (e.g., Tuinstra and Koenig, 1970; Jehlicka et al., 2003; Schopf et al., 2005). In perfectly crystalline graphite a single first order peak occurs at 1,582 $cm^{-1}$, attributed to stretching of the C–C bonds in basal graphite planes (known as the 'G' or 'graphite' peak). Second order peaks occur at ~2,695 $cm^{-1}$ and 2,735 $cm^{-1}$. Imperfectly crystallised graphitic carbons have additional peaks at ~1,355 $cm^{-1}$ (known as the 'D' or 'disordered' peak) and ~1,620 $cm^{-1}$ ('D'; occurring as a shoulder to the 'G' peak), and a single broad second order peak around 2,700 $cm^{-1}$. These peaks are also due to C–C bonds but arise because the lattice structure contains disorder. The exact position, height and width of these peaks can vary a little as a function of crystallite size and degree of ordering of the carbon. Raman spectral features have been used to try to extract information about both the biogenicity and antiquity of putative signs of early life. Antiquity has been investigated by comparing the change in spectral features of carbon to the change in thermal maturity of carbon through different metamorphic grades (e.g., Fig. A31; Tice et al., 2004; Wopenka and Pasteris, 1993). In general, both the Raman D/G peak area (i.e., the integrated intensity) ratio and the D/G peak width (i.e., the full peak width at half maximum intensity) ratio of the carbonaceous material decrease with progressive metamorphism. The D peak becomes narrower, sharper and of more prominent intensity, whilst the G peak shifts to lower wave numbers. The second order G peak also becomes more prominent.

Putative carbonaceous microfossils should exhibit the same Raman spectral features as other carbonaceous material in the same rock specimen because both should have undergone the same thermal maturation and geochemical alteration processes. Their spectra should also be consistent with known spectra from equivalent maximum metamorphic grade to the host rock unit, and should not contain evidence of hydrated minerals that would not survive metamorphism. However, care must be taken when making comparisons with carbon housed in rocks of equivalent metamorphic grade but from different geographical regions because the starting composition and behaviour of the carbon may have been different. Nevertheless, if serious inconsistencies exist, then it is likely that the 'microfossils' are younger contaminants.

While Raman spectra reveal both the nature of the molecular bonds (e.g., C–C) and the degree of crystallinity (order–disorder) of carbonaceous material, they tell the user nothing of the chemical composition (i.e., the ratio of elements). In complex biological carbonaceous compounds additional spectral peaks corresponding to bonds such as C = O, C–H and C–N might be expected. These have not yet been observed in ancient samples using laser Raman probably due to the degree of thermal/geochemical alteration. Indeed, similar Raman spectra are produced from synthesised non-biological disordered carbonaceous material as from *bona fide* biological kerogens (Fig. A32; Pasteris and Wopenka, 2003). Hence, Raman spectra cannot be used to

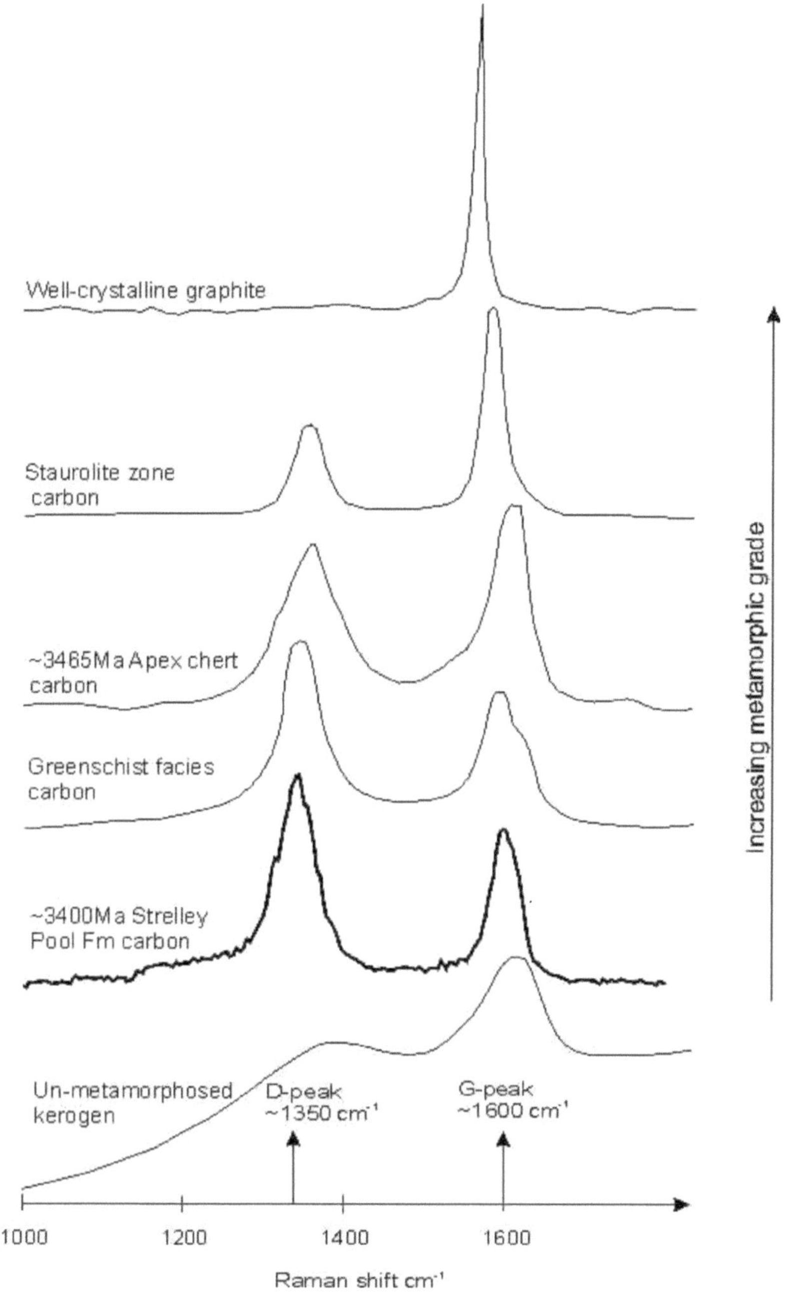


**Fig. A31** Representative Raman point spectra from carbon of different metamorphic grades (increasing grade from *bottom* to *top*) (Data from Brasier et al., 2002; Jehlicka et al., 2003; Pasteris and Wopenka, 2003). Spectra from carbon found in two rock units important in studies of early life (Apex chert and Strelley Pool Formation) are included for comparison. Both of these spectra show that the carbon has been through a metamorphic event consistent with the known metamorphic grade of the host rocks. This is a very useful technique to eliminate modern carbon contamination

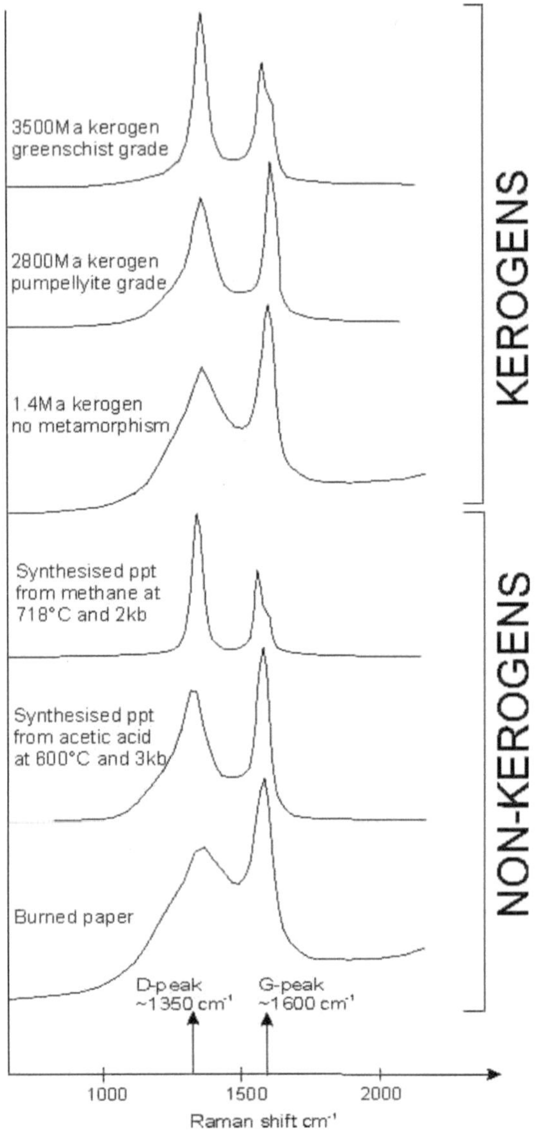


**Fig. A32** Raman spectra of biological (kerogen) versus non-biological material. Note the similarity between the 3,500 Ma greenschist grade kerogen and the synthesised non-biological methane precipitate; between the pumpellyite grade kerogen and the synthesised non-biological acetic acid precipitate; and between un-metamorphosed kerogen and burned paper. Clearly it is not possible for Raman to distinguish a non-biological from biological origin for carbonaceous material (Data from Pasteris and Wopenka, 2003)

*prove* the biogenicity of a putative microfossil or other carbonaceous material. However, Raman can be used as a useful elimination test for biogenicity – if a

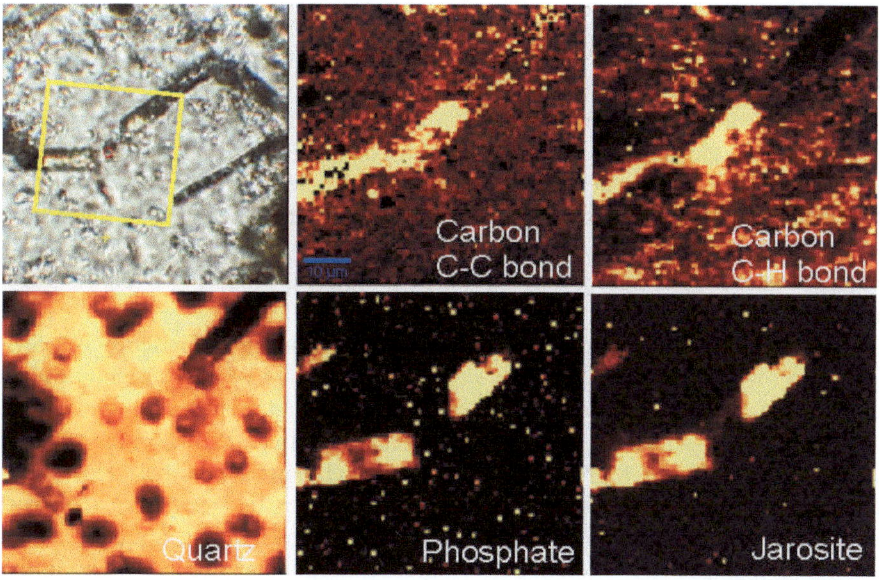


**Fig. A33** Raman images of a partly carbonaceous microtubular structure within the ~3,400 Ma Strelley Pool sandstone. Raman has detected both C–C and C–H bonds within the carbonaceous material. It also shows that the microtube is infilled with phosphate and jarosite. This evidence, especially the fact that jarosite is a hydrated mineral (O–H bonds) unlikely to survive metamorphism, suggests that this microtube is a younger contaminant, possibly formed by modern endolithic microbes (Images courtesy of Andrew Steele)

carbonaceous microstructure does not give the Raman spectra of structurally disordered carbon, then it can be disregarded for further study.

**Imaging**: Raman imaging allows the micro-mapping of the distribution of certain spectral parameters within a sample (Fig. A33). For example, the user might want to map how the ratio of D to G band varies over a given area, or which areas of a sample contain a certain mineral. 3-D maps are also obtainable by taking images at various depths within the thin section and then stacking them together using a 3-D software package (e.g., McKeegan et al., 2007). This allows a one-to-one correlation of carbonaceous molecular structural chemistry with optically recognisable morphological features. Most simply, this can be used to show the co-occurrence of biological morphology, such as putative cell walls, with disordered carbonaceous material (Kudryavtsev et al., 2001; Schopf and Kudryavtsev, 2005). Schopf et al. (2007) have done this using putative filamentous microfossils from the ~3,460 Ma Apex chert. They show two filamentous structures composed of hollow quartz filled 'cell cavities' with kerogen-like 'cell walls', closely comparable to younger *bona fide* Precambrian microfossils. Moving to even more ancient rocks, McKeegan et al. (2007) have used 3-D Raman imaging to demonstrate the existence of ~5 µm carbonaceous inclusions in apatite within >3,800 Ma rocks from Akilia Island, Greenland. Because of the non-destructive nature of the technique, they were able to follow this up by SIMS carbon isotope analysis of the same inclusions. Whilst Raman imaging does not provide *proof* of biogenicity in either of these cases, it is undoubtedly a very useful technique.

A recent additional application of Raman imaging is to look at the distribution and structure of the minerals which host purported biological material. Many minerals produce spectral peaks that vary in intensity depending on their crystallographic orientation to the laser. This feature can be used, for example, to image the distribution of quartz crystallographic axes (e.g., Fig. A33 bottom left) to see whether the biological material occurs between grain boundaries, is enclosed by entire grains, or occurs in cracks.

A closely related technique is **confocal laser scanning microscopy** (CLSM). CLSM allows rapid and precise *in situ* characterisation of 3D morphology and taphonomic preservation. It is also non-intrusive and non-destructive and its advantage over laser Raman for morphological investigations lies in the speed of acquisition and slightly higher spatial resolution (<100 nm is possible). However, it has currently only been applied to younger (micro)fossils freed from their rock matrix, or to Proterozoic permineralized microfossils in chert (Schopf et al., 2006). It relies on the laser excitation of auto-fluorescent organic material so faces limitations when objects are either a long way below the surface of the thin section or have thick opaque walls. More importantly, it cannot be used on samples which are composed of highly geochemically altered carbon which does not fluoresce. Nevertheless, CLSM may be a useful additional technique for selected samples from Pilbara and Barberton that have undergone only sub-greenschist facies metamorphism.

## 5.9 Near Edge X-Ray Absorption Fine Structure Spectroscopy (NEXAFS) and Electron Energy Loss Spectrometry (EELS)

These are two very similar techniques used primarily, in studies of early life, to investigate the nature of carbonaceous material. In particular, these techniques are used to look at chemical bonding and local atomic configuration. EELS uses electronic excitation to probe the structure of the carbonaceous material and is performed using an EELS spectrometer in a high resolution transmission electron microscope (HRTEM). NEXAFS (also known as XANES in some fields) uses x-rays instead of electrons to excite the atoms and is thus performed in a scanning transmission x-ray microscope (STXM). The source of the x-rays for very high resolution NEXAFS work is a synchrotron (see Section 5.10).

NEXAFS is highly sensitive to bond lengths and bond angles and is element specific (i.e., the user must know the target element before applying the technique). NEXAFS is usually used to investigate the chemical binding state of carbon atoms in both modern and ancient carbonaceous material, so spectra are recorded within

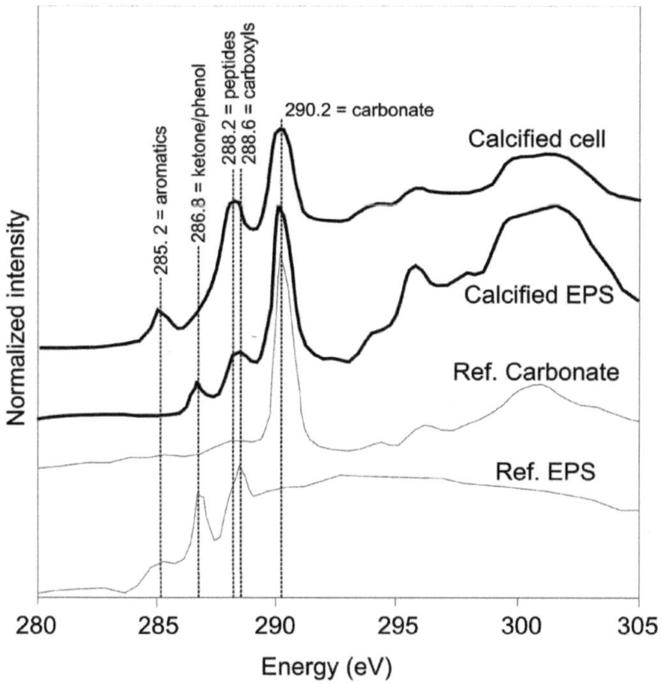


**Fig. A34** A typical NEXAFS spectrum showing how the chemical binding state of carbon atoms can be investigated. Here, spectra from a modern calcified cell and calcified EPS (*bold lines*) are compared to reference spectra from carbonate and EPS (*feint lines*). Carbon bound in aromatic, ketone, peptide, carboxyl and carbonate groups can be distinguished (Figure modified from Benzerara et al., 2006)

a specific energy range corresponding to the carbon K-edge (1s-shell electrons). Different absorption peaks can be seen for carbon bound, for example, in aromatic groups, ketones, peptides, carbonyls and carbonate (Fig. A34). This type of spectrum may help the user to distinguish between cellular and extra-cellular components and to strip out any signal from carbonate minerals. Spectral data can be combined with STXM images to build up maps of potential microfossils.

EELS measures the energy shift associated with the initial excitation of an atom by the electron beam to its excited or ionized state. The energy range of a typical EELS spectrum for carbon is 270–300 eV with a resolution of about 1 eV compared to about 0.1 eV for NEXAFS. Like NEXAFS, distinctive energy spectra reveal such structures as graphitic ordering, C = C double bonds (e.g., aromatic or alkenyl groups) and C = O bonds (e.g., carbonyl groups). In addition to the production of spectra, it is also possible to produce maps of these features on a near atomic scale.

These techniques can sometimes be used to distinguish between non-biological graphite and biological kerogenous material. However, some non-biological carbonaceous compounds made by, for example, Fischer Tropsch-type synthesis may contain similar functional groups to biological kerogen (e.g., De Gregorio et al., 2005). NEXAFS and EELS may therefore only give signals consistent with biology, rather than uniquely attributable to biology.

## 5.10 Synchrotron X-Ray Tomography

This is a non-intrusive, non-destructive technique for three dimensional compositional and morphological scans. X-rays are generated in a synchrotron when electrons are accelerated to giga electron volt (GeV) energies along a curved path within the synchrotron ring. The electron path is then bent by electromagnetic fields causing the electrons to decelerate and lose energy. The energy is lost in the form of photons with different wavelengths from infra-red all the way through to x-rays. The x-rays are collected and used as the 'light beam' for both this tomographic technique (and the STXM and NEXAFS techniques described above). Only synchrotron x-rays can produce the very small beam sizes needed for nano-scale work.

X-ray tomography has recently been successfully applied to the study of fossil embryos in the late Precambrian (Donoghue et al., 2006). Here, 3-D sub-micron scale images of fossils preserved as calcium phosphate were produced. Undoubtedly this technique will soon be applied in the Archean.

## 5.11 Atomic Force Microscopy (AFM)

AFM allows the user to investigate and image the structural organisation of the surface of fossilised material at the nanometer- to atomic-scale. This is achieved by 'feeling' the surface with a micro-scale cantilever and sharp tipped probe. Forces between the probe tip and the sample (e.g., electrostatic, chemical bonding, magnetic) lead to deflection of the cantilever and this deflection is measured by a laser. In its simplest operational mode, a feedback mechanism ensures a constant distance between the probe tip and the sample, and this effectively maps out the topography of the sample. Other measurement modes are possible dependent on the users needs.

AFM provides much of the same information as SEM but has several advantages. Samples require no coating or other special treatment that may prevent further analyses, no vacuum is required (this is particularly useful for biological samples and/or those in liquids), the AFM produces true 3-D images rather than 2-D projections of 3-D features, and typically the resolution is better. The SEM does, however, hold the advantage for objects with large depths of field.

AFM has been used to image organic walled fossils from the 650 Ma Chichkan Formation of Kazakhstan (Kempe et al., 2002), where ~200 nm platelets of polycyclic aromatic kerogen were identified as the structural units comprising the fossil walls. The orientation of these structural units may be used, in combination with other techniques, to deduce the biological versus non-biological nature of ancient carbonaceous material. A major drawback of AFM for work in the Archean is the need to expose a putative fossil wall exactly at the surface of a sample, whilst at the same time preserving its original ultra-structure; not a simple task!

## 5.12 Molecular Fossils

Molecular fossils or 'biomarkers' are essentially cell membrane lipids which can be preserved as soluble hydrocarbons in sediments; where sediments are sufficiently well preserved these hydrocarbons are thought to indicate the former presence of specific groups of organisms, such as cyanobacteria. As noted previously (Section 1.2), a *morphological* expression of the cell contents or cell membrane appears to have limited potential for preservation in the rock record. This does not appear to be the case for the *chemical* expression of such cellular components. Historically, it was thought that such complex hydrocarbons would not survive the metamorphism experienced by all Archean rocks; hence, if found, they could only be attributed to modern contamination. However, a re-assessment of the thermal stability of hydrocarbons (e.g., Mango, 1991) suggested that some could survive low grades of metamorphism.

Following this discovery, the pioneering work of Roger Summons and his team identified two types of molecular fossils from 2,700 Ma rocks from the lowermost Hamersley Group, Pilbara Craton, Western Australia (Brocks et al., 1999, 2003). The first of these are hopanoids, in particular 2a-methylhopane, which are claimed to be uniquely characteristic of aerobic cyanobacteria (but see below). If these biomarkers are truly syngenetic with the host rock, then this provides the first unequivocal evidence of cyanobacteria in the Archean. Likewise the identification of steranes, in particular cholestane, would testify to the presence of eukaryotes at 2,700 Ma, some 1,000 Ma earlier than previous evidence suggests. These are substantial and somewhat controversial claims and the evidence must be water tight.

Like all techniques in the Archean, molecular fossils come with their problems, not least contamination issues. For them to be truly useful it must be demonstrable without doubt that the molecules are actually Archean in age and indigenous to the samples in which they are found. This is perhaps even more difficult when working with molecular fossils than with microfossils or trace fossils. The molecular hydrocarbons are soluble, hence mobile, so the opportunity for contamination is immense, both during post-Archean geological time and during sample collection and preparation. Several lines of evidence, such as low levels of metamorphism, high contents of associated kerogen, covariance of $\delta^{13}C$ of kerogen and bitumen, remoteness of the sample from younger oil prone rocks, and distinctive carbon compound assemblages, have been used to indicate syngenecity of these biomarkers (Summons et al., 1999). The molecular fossils from the Hamersley Group have thus been described as 'probably syngenetic' with their host rocks (Brocks et al., 2003). We have to ask ourselves whether 'probably' is good enough to validate the significant claims for cyanobacteria, eukaryotes and an oxygenated atmosphere at 2,700 Ma – my view is 'probably not'. Further sampling and careful decoding of the various biomarker signals may however lead to these claims being substantiated.

The second difficulty with molecular fossils lies with the fact that only a small percentage of bacteria have so far been cultured. For example, can we be sure that hopanoids are only characteristic of cyanobacteria, or might some as yet uncultured bacteria also contain them? The first indication that this assumption does not hold came with the publication of data by Fischer et al. (2005). Here it was shown that certain other bacteria (in this case *Geobacter sulfurreducens*) produced a variety of complex hopanoids in strictly anaerobic conditions. Much more work therefore appears necessary to fully understand the significance of molecular fossils.

## 5.13 Carbon Isotopes

Carbon isotopes have traditionally been a central part of any claim for early life on Earth. The global carbon cycle involves exchange between three major reservoirs: gaseous carbon, mostly $CO_2$ and $CH_4$; carbonates; and organic carbon. While the two stable isotopes of carbon ($^{13}C$ and $^{12}C$) are present in the global system in fixed amounts, the ratio between them in a given reservoir of carbon may vary both spatially and with time. This ratio is controlled by thermodynamic and kinetic effects, with kinetic effects dominant in the low temperature biological environment. Here, the lighter isotope, $^{12}C$, is more easily captured and released from chemical bonds than the heavier isotope, $^{13}C$. Hence, when $^{13}C/^{12}C$ fractionation takes place during chemical reactions, the reservoir ratios will change from some central mean. Selective fixation of $^{13}C$ in one reservoir will therefore lower $^{13}C/^{12}C$ in the other and vice versa.

The ratio of $^{13}C/^{12}C$ is measured using a mass spectrometer and is expressed as:

$$\delta^{13}C_{\text{sample}}(‰) = [(^{13}C/^{12}C)_{\text{sample}}/(^{13}C/^{12}C)_{\text{standard}} - 1] \times 1{,}000 \quad (5.2)$$

quoted relative to the Peedee Belemnite (PDB) standard (Hoefs, 1987).

Two assumptions are inherent to the use of carbon isotopes as a tracer for early life:

- Early life was carbon based. This is a safe assumption; carbon-based life today must have evolved from a carbon-based ancestor, and no other elements share the properties of carbon that make it the ideal building block for life.
- Early life metabolisms exerted a fractionation on the isotopes of carbon. Experiments have shown that it is the initial carbon-fixing enzymatic carboxylation reaction in the metabolism of autotrophic organisms (Tables A1 and A2) that imparts the largest carbon isotope fractionation, due to the preferential biological uptake of $^{12}C$. Autotrophic organisms are those that take $CO_2$ from their environment and fix the inorganic carbon into biological end-products (as opposed to heterotrophs that derive their carbon from existing organic molecules; see Table A1). Logically, the first organisms would be autotrophic, since $CO_2$ was abundant and non-limiting in Archean environments. Hence, this second assumption also appears safe.

The *magnitude* of carbon isotope fractionations caused by biological carbon fixation pathways (Table 5.2) have been investigated in the laboratory and are fairly well constrained (Fig. A35 grey bars). A glance at Fig. A35 shows that the average

**Table A1** Classification of microbial metabolisms according to their carbon source and energy source. Evidence from both the rock record and from phylogenetic studies suggests that the earliest microbes were chemoautotrophs

| | |
|---|---|
| Photoautotrophs | • Carbon source: $CO_2$<br>• Energy source: light<br>• Examples: cyanobacteria (oxygenic), green and purple sulfur bacteria (anoxygenic). |
| Chemoautotrophs | • Carbon source: $CO_2$<br>• Energy source: oxidize inorganic compounds which are used to fix the $CO_2$<br>• Examples: nitrifying, hydrogen, sulfur and iron-utilizing bacteria; Archaea which live among hydrothermal ocean vents |
| Photoheterotrophs | • Carbon source: from organic compounds made by other organisms<br>• Energy source: light<br>• Examples: green and purple non-sulfur bacteria |
| Chemoheterotrophs | • Carbon source: from organic compounds made by other organisms<br>• Energy source: from oxidation of organic compounds<br>• Examples: most other bacteria |

**Table A2** Pathways for the autotrophic fixation of carbon into living biomass. The carbon isotope fractionations associated with common plant and microbe groups are given in Fig. A35

| Pathway | Summary reaction | Examples |
|---|---|---|
| C3 | $CO_2$ + ribulose-1,5-biphosphate → phosphoglycerate | Green C3 (compound with a 3-carbon skeleton) plants; eukaryotic algae; cyanobacteria; purple photosynthetic bacteria; purple non-sulphur bacteria |
| C4 and CAM (crassulacean acid metabolism) | $CO_2/HCO_3^-$+ phosphoenolpyruvate/ pyruvate → oxaloacetate | Green C4 and CAM plants (combined with C3 reaction); some anaerobic bacteria; facultatively aerobic bacteria |
| Reverse TCA (tri-carboxylic acid) | $CO_2$ + succinyl CoA → acetyl CoA + pyruvate | ε-Proteobacteria; green sulphur bacteria; thermophilic hydrogen oxidising bacteria; some Archaea; some sulphate reducing bacteria |
| Reductive acetyl coenzyme A | $CO_2$ + acetyl CoA → pyruvate | Some sulphate-reducing bacteria; some methanogenic bacteria; *Clostridium* |
| 3-Hydroxypropionate | $CO_2$ + acetyl CoA and propionyl CoA → glyoxylate | Green non-sulphur photosynthetic bacteria (*Chloroflexus*); some Archaea |
| Ribulose monophosphate | $CH_4$ → formaldehyde (HCHO) followed by HCHO + ribulose monophosphate → hexulose monophosphate | Type I methanotrophic bacteria |
| Serine | $CH_4$ → formaldehyde (HCHO) followed by HCHO + glycine → serine | Type II methanotrophic bacteria |

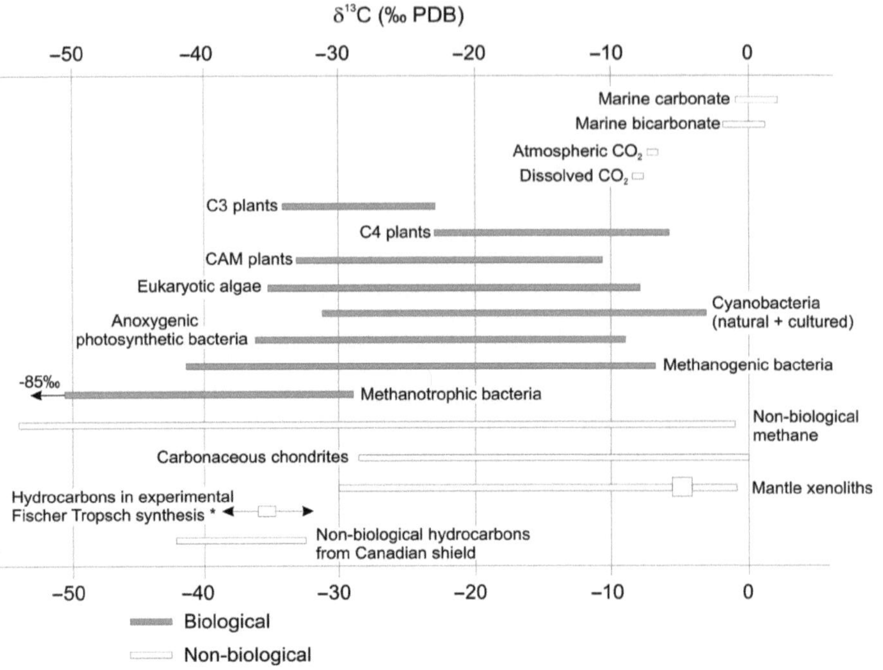


**Fig. A35** Carbon isotope composition of the various biological and non-biological carbon reservoirs. It is clear that non-biological carbonaceous compounds may possess $\delta^{13}C$ values that span the same range of isotopic compositions as compounds formed from biological reactions (Data is from Horita and Berndt, 1999; Schidlowski, 2001; Sherwood Lollar et al., 2002; Krot et al., 2003; McCollom and Seewald, 2006. *Note that the hydrocarbons in this experiment were depleted by 36‰ compared to the reactant $CO_2$; their absolute $\delta^{13}C$ value was ~–50‰ PDB)

$\delta^{13}C$ value of the global biomass (grey bars) is around 20–30‰ more negative than the dominant inorganic carbon species (i.e., marine bicarbonate) in the surface environment. With this knowledge at hand, a first order assessment of whether or not biological fractionation occurred in the past can be made by looking for similar discrepancies in the sedimentary rock record.

Carbonate sediments have been widely used as a proxy for the $\delta^{13}C$ of the *inorganic* pool through time. When carbonates precipitate in the oceans the carbon is derived from the dissolved inorganic carbon (DIC) pool. This pool comprises $CO_2$ (aq), $HCO_3^-$ (aq) and $CO_3^{2-}$ (aq). Since the carbon isotope fractionation between DIC and $C_{carbonate}$ is negligible (about 1‰; Morse and McKenzie, 2002) the $\delta^{13}C$ of ancient carbonates should track that of DIC. The $\delta^{13}C$ of the *biosphere* through time has been measured directly from carbonaceous material (kerogen) found in ancient sediments. Primary biological matter undergoes multiple stages of transformation during diagenesis with kerogen being the end-product. Kerogen is characterized as a chemically inert, acid-insoluble aggregate of aliphatic and aromatic hydrocarbons (Durand, 1980). The primary isotopic composition of this kerogen may, however, be affected by contamination by younger externally derived hydrocarbons and by metamorphism (and associated conversion to graphite). Hence, great care must be taken to find the least altered kerogen which is truly representative of the biosphere at the time of interest (e.g., Des Marais et al., 1992).

Given the fact that there is a continuous record of carbonate carbon and organic carbon back to the earliest sedimentary rocks (~3,800 Ma) it is not surprising that carbon isotopes have been one of the most popular tools for studying early life. The common occurrence in the Archean of organic carbon with $\delta^{13}C$ values displaced by −20‰ or more from carbonate carbon (0 ± 5‰ PDB) (see Schidlowski, 2001) has led many researchers to claim biological activity throughout the entire sedimentary rock record, right back to ~3,800 Ma. Indeed, the *magnitude* of $^{13}C$ depletion in putatively kerogenous material has even been used by some workers to suggest the *type* of microbes producing the fractionation. For example, values of −20‰ to −30‰ (PDB) have been taken to indicate the presence of photosynthetic microbes (e.g., Tice and Lowe, 2004), whereas lighter values (e.g., −56‰ PDB; Ueno et al., 2006) have been explained by the activity of methanogenic microbes. These $\delta^{13}C$ values are certainly consistent with life but, unfortunately, this may be somewhat of a simplification of events in the early Archean.

It must be highlighted that carbon isotope fractionation is by no means a uniquely biological signal (Fig. A35, white bars). A 'null hypothesis' that needs to be falsified for each potentially biological signal is the alternative non-biological origin for the light carbon, for example from Fischer–Tropsch type (FTT) reactions between CO and metals (Horita and Berndt, 1999; Sherwood Lollar et al., 2002) and/or from the metamorphic reduction of siderite (van Zuilen et al., 2003). These can both result in carbon isotope fractionations that lie within the 'biological domain'. A non-biological source for some of this Archean carbon is supported by the ubiquity of light carbon in deep hydrothermal vein cherts, and by close association with hydrothermal carbonates, sulphates and metals (Brasier et al., 2002, 2005; Lindsay et al., 2005). This scenario, that there was deep, abundant, hydrothermally-generated non-biological carbon like that found around some modern black smokers, urgently needs to be further tested. In addition, meteorites (Sephton and Gilmour, 2001) and organic haze (Pavlov et al., 2001) can generate suites of non-biological organic compounds with negative carbon isotope signatures.

A recent report (Nemchin et al., 2008) uses carbon isotopes to raise the tantalising possibility that life may have been in existence prior to the time that we have a preserved rock record. This study reports carbon isotope values with a range of −5‰ to −58‰ PDB from tiny slivers of diamond and graphite within detrital zircons ($ZrSiO_4$) from Jack Hills in Western Australia. These zircons have been dated at ~4,250 Ma; this would push the appearance of life on Earth back another 500 Ma from the next oldest claim from Greenland (which is also questionable). Many researchers are sceptical, including the authors of the study, who rightly state that other non-biological processes could have induced similar isotopic fractionations. Others go further, suggesting that the diamond slivers are contamination occurring during preparation of the sample (samples are often cut and polished using diamond based products). The particularly wide spread of carbon isotopes values has also been taken by some to indicate a non-biological reaction sequence rather than a biological source, which may be expected to be less variable. The biggest problem with this claim is the fact that the host rock has not been preserved, we only have individual zircon crystals. This means there is no way of knowing the true geological context of the putative biological signal. There also needs to be some mechanism to transport 'biological carbon', which presumably existed somewhere near the surface of the Earth, to the great depths and pressure for it to be transformed into diamond. This would imply that plate tectonics and subduction zones were already in existence, another controversial claim. While this claim is very interesting, it seems that a non-biological fractionation mechanism is much more likely to explain the quoted $\delta^{13}C$ values.

For reasons highlighted above, the trust placed in the earliest carbon isotope record as a *single* line of evidence for life has in the eyes of most geologists now outrun its term. Carbon isotopes are certainly one necessary and very valuable technique to investigate carbonaceous matter in Earth's oldest rocks. They will, however, only ever give evidence that is consistent with life, rather than unique to life. In this early rock record carbon isotopes must, if possible, be integrated with other isotope systems that purport to reflect biological fractionation (see below), and with a complete understanding of the geological context and biological fractionation pathways.

A final note on carbon isotopes concerns the instrumentation available to measure $^{13}C/^{12}C$ ratios. Most $\delta^{13}C$ data, especially prior to about 2000, had been obtained using bulk analyses of carbonaceous rocks, whereby bulk organics are isolated from a rock, combusted, then the $CO_2$ produced is analysed for its isotopic composition. With recent advances in mass spectrometry, *in situ* measurements can be obtained from thin sections containing individual putative fossils and/or carbonaceous laminations using SIMS (e.g., Ueno et al., 2001) or nanoSIMS (e.g., Wacey et al., 2008). *In situ* measurements are much more reliable than bulk measurements because the possibility of contamination from other carbon containing compounds in the bulk sample is removed.

## 5.14 Sulphur Isotopes

Sulphur isotopes appear, at present, to be the most useful isotopic system for investigating early biosphere and atmosphere processes. There is a general consensus that primitive bacteria which metabolise sulphur compounds are one of the most deeply rooted groups in the 'Tree of Life' (e.g., Shen et al., 2001). Sulphur isotope analyses of ancient sulphides (mostly pyrite) and sulphates are being used in an attempt recognise processes such as biological sulphur oxidation, sulphate reduction and sulphur disproportionation. Sulphur isotopes have also provided a hotly debated tracer for the rise of atmospheric oxygen, an application that will not be discussed further here (see for example, Farquhar et al., 2000; Holland, 2006; Kasting and Ono, 2006).

The ratio of $^{34}S/^{32}S$ preserved in sedimentary sulphides is a well established geochemical tool to detect the activity of sulphur-metabolising microbes in the rock record. Traditionally, this ratio, expressed as:

$$\delta^{34}S_{sample}(‰) = [(^{34}S/^{32}S)_{sample}/(^{34}S/^{32})_{standard} - 1] \times 1{,}000 \quad (5.3)$$

relative to the Canyon Diablo Troilite (CDT, e.g., Hoefs, 1987), has been used to detect the presence of sulphate-reducing bacteria. During microbial sulphate reduction (MSR), the lighter $^{32}S$ ions react more frequently than $^{34}S$ ions, so that lighter S-isotopes become preferentially incorporated into $H_2S$ and sequestered into solid sulphide. Both pure cultures and natural populations of sulphate-reducing microbes produce sulphides with $^{34}S$ depletions of up to ~45‰ when compared with co-existing sulphate (Bollinger et al., 2001; Habicht and Canfield, 2001), while theoretical calculations suggest that potential depletions could reach ~70‰ (Brunner and Bernasconi, 2005). *In situ* measurements of hyper-sulphidic sediments have revealed $^{34}S$ depletions of up to 77‰ (Ruchniki et al., 2001) but the extent to which this is attributed to sulphate reduction alone is currently unknown. At very low sulphate concentrations (<50–200 μM) microbial fractionation becomes negligibly small (Habicht et al., 2002). S-isotope fractionation also occurs during the microbial reduction and disproportionation of sulphur species such as elemental sulphur ($S^0$), thiosulphate ($S_2O_3^{2-}$) and sulphite ($SO_3^{2-}$) (Canfield and Thamdrup, 1994; Canfield et al., 1998).

Interactions in the sulphur isotope system are complex and result in a wide range of isotopic compositions of sedimentary sulphides, sulphates and intermediate sulphur species from about −65‰ to +90‰ (Hoefs, 1987). From the Neoproterozoic onwards observed 50–60‰ fractionations (see Canfield and Raiswell, 1999) between sulphides (depleted in $^{34}S$) and co-existing sulphates are attributed to MSR. In older rocks such fractionations are much reduced. Most sedimentary sulphides older than ~2,700 Ma have narrow S-isotopic ranges ($\delta^{34}S$ = 0‰ ± 5‰). This has led to contrasting interpretations for the process of pyrite formation at this time: possibly from $H_2S$ derived non-biologically from hydrothermal or volcanogenic processes (implying that before 2,700 Ma SRB had not evolved, the oceans were largely free of sulphate and the atmosphere contained virtually no free $O_2$); possibly from MSR in oceans with little sulphate (<<1 mM) causing minimal isotopic fractionation; or possibly from rapid MSR in a warm ocean with abundant sulphate (e.g., Ohmoto and Felder, 1987), although laboratory experiments raise questions about the latter (Habicht and Canfield, 1996).

The largest $\delta^{34}S$ fractionations (maximum ~22‰; mean ~12‰) reported from early Archean rocks come from co-occurring sulphide and barite minerals in the ~3,490 Ma Dresser Formation of Western Australia (Shen et al., 2001; Philippot et al., 2007), and have been argued to result from MSR (Shen et al., 2001) or microbial sulphur disproportionation (Philippot et al., 2007). Unfortunately, this barite may be a primary hydrothermal precipitate (Runnegar et al., 2001) rather than a replacement of low temperature sedimentary gypsum, as originally assumed. Since even the maximum fractionations of around 22‰ are within the range observed in non-biological thermochemical reactions, these $\delta^{34}S$ data remain equivocal.

Sulphur isotopes give more definitive evidence for the emergence of sulphur processing microbes by the late Archean. The 2,700–2,600 Ma Belingwe Belt of Zimbabwe is one of the least metamorphosed of all Archean greenstone successions.

Here, analysis of pyrites from sulphidic shales of the Manjeri formation, in particular, reveal a wide range of $\delta^{34}S$ values (−21.1‰ to +16.7‰). This range, together with the pyrite morphology and heterogeneity of isotopic compositions within single pyrites, plus supporting carbon isotope data, provides strong evidence for sulphate reducing and possible sulphur oxidising bacteria at this time (Grassineau et al., 2002).

It is still somewhat uncertain whether sulphur based metabolisms originated very early in Earth's history, in response to the availability of sulphur species from volcanic vents and from anoxygenic photosynthesis, or later, as a response to the changing redox status of the planet. The ability to measure sulphur isotopes, in combination with other isotopic systems (*in situ*) on individual sulphide grains of smaller dimensions (for example using SIMS and nanoSIMS) may soon resolve this debate once and for all.

## 5.15 Other Isotopic Systems

**Nitrogen** has been an essential biological element since the emergence of life. Indeed, reduced nitrogen compounds such as ammonia must have been used in the *prebiotic* synthesis of amino acids and proteins. Biological nitrogen fixation is believed to have evolved very early in the history of life; phylogenetic studies suggest that nitrogen fixation occurred before the divergence of the Bacterial and Archaeal domains (e.g., Young, 1992). Hence, nitrogen can be used as a potential biosignature throughout the rock record.

Fractionations of the two most abundant isotopes of nitrogen ($^{15}N/^{14}N$) are used to investigate reactions in the modern nitrogen cycle. Nitrogen isotopes are expressed using the delta notation:

$$\delta^{15}N_{sample}(‰) = [(^{15}N/^{14}N)_{sample}/(^{15}N/^{14}N)_{standard} - 1] \times 1{,}000 \qquad (5.4)$$

quoted relative to atmospheric nitrogen (AIR).

In the modern Nitrogen cycle, the major source of $\delta^{15}N$ fractionation is biological, occurring during the dissimilatory denitrification of nitrate ($NO_3^-$). This results in residual oceanic $NO_3^-$ being enriched in $^{15}N$ by about 6–7‰ relative to atmospheric nitrogen (Boyd, 2001 and references therein). Subsequently, this $^{15}N$ enriched nitrate can be reduced to $NH_4^+$ and assimilated by organisms. This heavy (+6‰AIR to +7‰AIR) signal is passed into the geological record in sedimentary organic matter. The $\delta^{15}N$ signal has been shown to be quite durable and survive moderate metamorphism with only a little alteration (e.g., enrichment of $^{15}N$ of only 1–2‰ from sub-greenschist to lower amphibolite facies; Jia, 2006). Other biological reactions can also be tracked by $\delta^{15}N$. For example, chemotrophic reactions such as direct fixation of $N_2$ or direct assimilation of $NH_4^+$ which result in negative fractionations (up to about −9‰; Delwiche and Steyn, 1970; Goericke et al., 1994). In addition, a biological $\delta^{15}N$ signal may be preserved in $NH_4^+$ minerals; degradation of organic material during diagenesis releases $NH_4^+$ (with little apparent $\delta^{15}N$ fractionation; Williams et al., 1995); this has a high thermal stability when incorporated in silicate minerals such as micas and feldspars during metamorphism, and could potentially be present as a biosignature when other traces of life have been destroyed (Boyd, 2001). Indeed, Papineau et al. (2005) have reported $\delta^{15}N$ values of ~−2‰AIR to +6‰AIR from ~3,800 Ma $NH_4^+$ rich biotite from the Isua Supracrustal Belt in south-western Greenland, claiming an early evolution of nitrogen fixation and/or ammonium assimilation.

The change in $\delta^{15}N$ through time in sedimentary rocks has thus been used to gain insights into the evolution of the biosphere and hydrosphere. $\delta^{15}N$ values from kerogenous cherts in various Archean greenstone belts (Beaumont and Robert, 1999) are rather negative (0‰AIR to −6‰AIR) compared to Proterozoic and Phanerozoic examples (0‰AIR to +10‰AIR). Since the positive $\delta^{15}N$ enrichments in modern sedimentary organic mater are due to the $^{15}N$ enrichment in $NO_3^-$ (see above), this data has been taken to imply that $NO_3^-$ was absent in Archean oceans, the oceans were largely anoxic and any microorganisms were using reduced nitrogen forms ($N_2$, $NH_3$ and $NH_4^+$) in chemo(litho)trophic metabolisms.

Two questions must be asked of currently available Archean nitrogen isotope data in relation to the earliest biosphere:

1. Do the $\delta^{15}N$ values reported represent global processes or just very localized fractionation around, for example, hydrothermal vents?

It is interesting to note that $\delta^{15}N$ of Archean kerogens is rather similar to $\delta^{15}N$ values observed in modern sub-seafloor chemotrophic ecosystems in the vicinity of hydrothermal vents. The reported $\delta^{15}N$ of this hydrothermal biota is −12‰AIR to +4‰AIR (e.g., Rau, 1981).The geological settings of many of the cherts described by Beaumont and Robert (1999) are consistent with such a chemotrophic hydrothermal community. This conclusion is further supported by the work of Pinti et al. (2001) who report a $\delta^{15}N$ of −7.4‰AIR for kerogenous chert from North Pole in the Pilbara of Western Australia. Pinti et al. (2001) conclude that this is a pristine Archean value and that it represents a 'metabolic isotopic fractionation induced by chemotrophic bacteria using $NH_4^+$ contained in hydrothermal fluids'. In addition, Pinti et al. (2007) report $\delta^{15}N$ values of about +6‰AIR to +7‰AIR from Fe–Mn oxyhydroxides within the ~3,460 Ma Marble Bar chert from Western Australia. The similarity of these values to Proterozoic and Phanerozoic sedimentary organic matter led them to conclude that there may have been a localized oxygenated environment allowing similar biochemical nitrogen cycling to today (the possibility of later nitrogen incorporation/contamination was also acknowledged). Given these conclusions, $\delta^{15}N$ values in Archean kerogens may not yet be informative about the redox state of the hydrosphere, they may simply be due to sampling bias.

2. Is there a uniquely biological $\delta^{15}N$ signal?

The short answer is no. The use of nitrogen isotopes as a biosignature is still in its infancy and there is still much to be learnt about potential fractionations and reservoir mixing on the earliest Earth. For this reason, nitrogen isotopes must be combined with other lines of evidence in any claim for early life.

Data from several late Archean greenstone belts reveal very positive $\delta^{15}N$ values for Archean nitrogen in kerogenous shales (>+10‰AIR; Kerrich et al., 2006).

These are $^{15}N$ enriched compared to Phanerozoic equivalents. When combined with previous analyses this gives a total range for Archean nitrogen hosted in cherts and shales of about −6‰AIR to +30‰AIR. Both Kerrich et al. (2006) and Jia (2006) attribute this wide range of Archean values to mixing between a mantle $N_2$ source (−5‰AIR) and a marine sedimentary kerogen component inherited from CI chondritic material (+35‰AIR to +42‰AIR; Kerridge, 1985). This suggests that N-isotopes do not provide a robust signal for either the redox-state of the planet, nor for tracing particular metabolic pathways.

Of great relevance to studies of early life, van Zuilen et al. (2005) report $\delta^{15}N$ values from graphite in ~3,800 Ma Isua supracrustal rocks. $\delta^{15}N$ values of around −3‰AIR to −1‰AIR were reported for graphite thought to have formed non-biologically through the thermal disproportion of Mg–Mn-siderite (van Zuilen et al., 2002). While these values are within the range of $\delta^{15}N$ found in chemotrophic hydrothermal vent biota (see above), they are also of similar magnitude to nitrogen found in magmatic fluids and modern mid-ocean-ridge-basalts ($\delta^{15}N \sim -5$‰AIR). Indeed, allowing for the slight enrichment in $^{15}N$ due to metamorphism (typically 1–2‰; Jia, 2006), the pristine values are probably very close to −5‰AIR. The geological setting dictates that the source of nitrogen was probably due to leaching of basaltic rocks by metosomatic fluids, so the negative $\delta^{15}N$ signature in these rocks does not constitute a biosignature. The realisation that nitrogen can also be introduced by diffusion or by metamorphic fluids at a later stage but still be strongly bound in a graphite structure (van Zuilen et al., 2005) makes the interpretation of nitrogen isotopes as a potential biosignature extremely difficult.

The last decade has seen the growing use of non-traditional heavy stable isotopes at the biology – geology interface (e.g., Iron, Nickel, Molybdenum). This has been due in part to increased understanding of the biological processing of transition and other 'heavy' elements, and in part to rapid advances in isotopic analytical techniques.

**Iron** isotopes are of particular interest in studies of early life because iron is essential for almost all organisms, mediating electron transfer in many enzymes, as well as acting as both electron donor and electron acceptor in primitive microbial metabolisms (Johnson and Beard, 2005). Its presence in magnetite is also of interest because, in some cases, this mineral is considered a biomarker. Iron is the most abundant element on our planet to participate in redox reactions; fractionation of its two most common stable isotopes ($^{54}Fe$ and $^{56}Fe$) during such reactions may reveal important information about the formation mechanisms of early Earth banded-iron-formations (e.g., Dauphas et al., 2007; Johnson et al., 2007; Whitehouse and Fedo, 2007), the redox balance of the early Earth (e.g., Rouxel et al., 2005), and the evolution of metabolic pathways involving iron oxidation and iron reduction (e.g., Beard et al., 1999). One specific application, relevant here, may be in tracing the evolution of dissimilatory $Fe^{3+}$ reduction (DIR). It has been suggested, using evidence from molecular phylogeny, that DIR may be one of the most ancient metabolic pathways (Vargas et al., 1998), perhaps more ancient than sulphate or nitrate reduction; the study of iron isotopes may eventually calibrate this important node on the 'Tree of Life'.

Most iron isotope geochemistry has so far concentrated on constraining the $\delta^{56}Fe$ of the various potential sources of iron on the early Earth, and experimentally determining fractionation factors between reduced and oxidised iron in biological and non-biological reactions. The value of iron isotopes as a biomarker in ancient rocks is still under debate, but several important data sets have already emerged (see reviews in Anbar, 2004 and Dauphas et al., 2007). Igneous rocks have very homogenous $\delta^{56}Fe$ values (±0.1‰), and oceanic and continental crust $\delta^{56}Fe$ are largely indistinguishable from that of bulk Earth. Modern terrestrial weathering, transport and deposition also results in negligible fractionations in bulk $\delta^{56}Fe$, although some rivers contain isotopically light dissolved iron (−1‰). The same appears to be true for terrestrial weathering at least as far back as 3,250 Ma. In contrast, marine sedimentary rocks that have undergone anoxic diagenesis appear to have largely negative $\delta^{56}Fe$ values. The total spread of $\delta^{56}Fe$ in material from low temperature sedimentary environments analysed thus far is about 4‰, and Precambrian sedimentary rocks contain iron bearing minerals that span almost this entire $\delta^{56}Fe$ range.

In laboratory experiments, iron redox reactions result in preferential enrichment of the heavy isotope in the oxidised ($Fe^{3+}$) form. This fractionation between $Fe^{2+}$ and $Fe^{3+}$ can be as much as 2–4‰ and appears to be independent of the redox mechanism. For example, $Fe^{2+}$ produced biologically by DIR was fractionated by −1.3‰ (Beard et al., 1999) relative to its $Fe^{3+}$ source, and $Fe^{3+}$ produced biologically by oxidation of $Fe^{2+}$ by anaerobic photoautotrophic bacteria was fractionated by +1.5‰ (Croal et al., 2004). However, a similar magnitude of fractionation (0.9–1.8‰) was observed during the non-biological precipitation of ferrihydrite ($Fe^{3+}$) from dissolved $Fe^{2+}$ (Bullen et al., 2001). This similarity of fractionation factors during biological and non-biological redox reactions makes the interpretation of any Fe-isotope signal in ancient rocks very difficult. Indeed, laboratory experiments (Balci et al., 2006) have found that isotopic fractionations during microbially-mediated Fe-reactions were primarily controlled by non-biological kinetic and equilibrium factors. This hurdle may not, however, be any greater than that faced by carbon or sulphur isotope geochemists. With these isotopic systems, even after many decades of study, $\delta^{13}C$ and $\delta^{34}S$ data only rarely give indisputable evidence of biogenicity. In most cases, they merely provide evidence consistent with biogenicity; it is then up to the researcher to analyse the geological and petrographic context of the samples to deduce whether non-biological mechanisms mimicking biological fractionations can be rejected.

The way forward for Fe-isotope biogeochemistry appears to be the integration of Fe-isotopes with detailed petrography, geological mapping, and other isotope systems, so that biological fractionations can truly be separated from non-biological ones. The first studies of this kind on rocks of Archean age are now being reported. For example, Yamaguchi et al. (2005, 2007) have combined Fe-isotopes with S-isotopes and elemental geochemistry to suggest that there was a diverse microbial community in the late Archean, including cyanobacteria, sulphate-reducing bacteria, methanogens and dissimilatory iron-reducing bacteria. Similarly, Archer and Vance (2006) have used the covariance of light Fe- and S-isotopes from 2,700 Ma pyrites to argue for sulphate- and iron-reduction metabolisms at this time. Van Zuilen et al. (2007) have combined Fe-isotopes with

**Table A3** Summary of common techniques used in studies of early life

| Technique | Main uses for early life studies |
|---|---|
| Geological mapping | Investigates the geological context of the host rocks; determines if context is plausible for life; determines age relationships with other rock units |
| Radiometric dating | Provides absolute ages for host rocks or constrains host rocks between two known ages |
| Petrography | Examines the morphology of the putative biosignature; helps to constrain the age of said structure with respect to other fabrics in the rock |
| Scanning electron microscopy (SEM) | Examines the surface morphology and texture of putative biosignatures; helps to constrain the age of said structure with respect to other fabrics within the rock |
| Transmission electron microscopy (TEM) | Investigates the crystal structure of minerals, especially useful for carbon; produces nano-scale chemical maps |
| Secondary ion mass spectrometry (SIMS) | Enables in situ elemental mapping and isotopic measurements |
| Nano-scale secondary ion mass spectrometry (NanoSIMS) | Enables in situ elemental mapping and isotopic measurements on the nano-scale. 10–50 times better than conventional SIMS |
| Elemental abundances and ratios (e.g., H/C) | Detects groups of elements that are important in biological reactions; helps to constrain the amount of alteration of samples |
| Laser Raman micro-spectroscopy | Characterises the bonding and crystallinity in carbonaceous compounds; helps to constrain syngenicity and allows elimination of distinctive non-biological carbonaceous material |
| Confocal laser scanning microscopy | Non-destructive 2-D and 3-D morphological analysis |
| EELS and/or NEXAFS | Characterisation of carbonaceous material at the nano- or atomic-scale |
| Synchrotron x-ray tomography | Non-destructive 2-D and 3-D morphological and chemical analysis |
| Atomic force microscopy (AFM) | Characterises surface structure on the nano- to atomic-scale |
| Molecular fossils (biomarkers) | Provides evidence for the former presence of particular groups of organisms (e.g., cyanobacteria, eukaryotes) |
| Carbon/sulphur/iron/nitrogen isotopes | Detects potential biological processing of these elements |

C-isotopes and detailed geological and petrographic mapping to invoke a biological photosynthetic iron-oxidation mechanism for iron oxides in the ~3,490 Ma Dresser Formation of Western Australia.

Further studies of the isotopic behaviour of iron in natural aqueous and biological samples, in ancient rocks, in laboratory experiments, together with theoretical calculations are all still necessary and represent exiting potential research fields for the future.

# Recommended Reading


Anbar, A. D., 2004, Iron stable isotopes: beyond biosignatures, *Earth Planetary Science Letters* **217**: 223–236.

Archer, C., and Vance, D., 2006, Coupled Fe and S isotope evidence for Archean microbial Fe(III) and sulfate reduction, *Geology* **34**: 153–156.

Balci, N., Bullen, T. D., Witte-Lien, K., Shanks, W. C., Motelica, M., and Mandernack, K. W., 2006, Iron isotope fractionation during microbially stimulated Fe(II) oxidation and Fe(III) precipitation, *Geochimica et Cosmochimica Acta* **70**: 622–639.

Beard, B. L., Johnson, C. M., Cox, L., Sun, H., Nealson, K. H., and Aguilar, C., 1999, Iron isotope biosignatures, *Science* **285**: 1889–1892.

Beaumont, V., and Robert, F., 1999, Nitrogen isotope ratios of kerogens in Precambrian cherts: a record of the evolution of atmosphere chemistry? *Precambrian Research* **96**: 62–82.

Benzerara, K., Menguy, N., Lopez-Garcia, P., Yoon, T-H., Kazmierczak, J., Tyliszczak, T., Guyot, F., and Brown Jr., G. E. D., 2006, Nanoscale detection of organic signatures in carbonate microbialites, *PNAS* **103**: 9440–9445.

Bollinger, C., Schroth, M. H., Bernasconi, S. M., Kleikemper, J., and Zeyer, J., 2001, Sulfur isotope fractionation during microbial sulfate reduction by toluene-degrading bacteria, *Geochimica et Cosmochimica Acta* **65**: 3289–3298.

Boyd, S. R., 2001, Ammonium as a biomarker in Precambrian metasediments, *Precambrian Research* **108**: 159–173.

Brasier, M. D., Green, O. R., Jephcoat, A. P., Kleppe, A. K., Van Kranendonk, M. J., Lindsay, J. F., Steele, A., and Grassineau, N. V., 2002, Questioning the evidence for Earth's oldest fossils, *Nature* **416**: 76–81.

Brasier, M. D., Green, O. R., Lindsay, J. F., McLoughlin, N., Steele, A., and Stoakes, C., 2005, Critical testing of Earth's oldest putative fossil assemblage from the ~3.5 Ga Apex Chert, Chinaman Creek, Western Australia, *Precambrian Research* **140**: 55–102.

Brasier, M. D., Green, O. R., Lindsay, J. F., McLoughlin, N., Stoakes, C. A., Brasier, A., and Wacey, D., 2008, Earth's oldest putative fossil assemblage from the ~3.5Ga Apex chert, Chinaman Creek, Western Australia: a field and petrographic guide, *Special Publication of the Geological Survey of Western Australia*.

Brocks, J. J., Logan, G. A., Buick, R., and Summons, R. E., 1999, Archean molecular fossils and the early rise of eukaryotes, *Science* **285**: 1033–1036.

Brocks, J. J., Buick, R., Logan, G. A., and Summons, R. E., 2003, Composition and syngeneity of molecular fossils from the 2.78 to 2.45 billion-year-old Mount Bruce Supergroup, Pilbara Craton, Western Australia, *Geochimica et Cosmochimica Acta* **67**: 4289–4319.

Brooks, J., Muir, M. D., and Shaw, G., 1973, Chemistry and morphology of Precambrian microorganisms, *Nature* **244**: 215–217.

Brunner, B., and Bernasconi, S. M., 2005, A revised fractionation model for dissimilatory sulfate reduction in sulfate reducing bacteria, *Geochimica et Cosmochimica Acta* **69**: 4759–4771.

Bullen, T. D., White, A. F., Childs, C. W., Vivit, D. V., and Schulz, M. S., 2001, Demonstration of significant abiotic iron isotope fractionation in nature, *Geology* **29**: 699–702.

Canfield, D. E., and Raiswell, R., 1999, The evolution of the sulphur cycle, *American Journal of Science* **299**: 697–723.

Canfield, D. E., and Thamdrup, B., 1994, The production of 34S-depleted sulfide during bacterial disproportionation of elemental sulfur, *Science* **266**: 1973–1975.

Canfield, D. E., Thamdrup, B., and Fleischer, S., 1998, Isotope fractionation and sulfur metabolism by pure and enrichment cultures of elemental sulfur-disproportionating bacteria, *Limnology and Oceanography* **43**: 253–264.

Croal, L. R., Johnson, C. M., Beard, B. L., and Newman, D. K., 2004, Iron isotope fractionation by Fe(II)-oxidizing photoautotrophic bacteria, *Geochimica et Cosmochimica Acta* **68**: 1227–1242.

Dauphas, N., Cates, N. L., Mojzsis, S. J., and Busigny, V., 2007, Identification of chemical sedimentary protoliths using iron isotopes in the > 3750 Ma Nuvvuagittuq supracrustal belt, Canada, *Earth Planetary Science Letters* **254**: 358–376.

De Gregorio, B., T., Sharp, T. G., and Flynn, G. J., 2005, A comparison of the structure and bonding of carbon in Apex chert kerogenous material and fischer-tropsch-type carbons, *Lunar and Planetary Science* **XXXVI**: A1866.

Delwiche, C. C., and Steyn, P. L., 1970, Nitrogen isotopic fractionation in soils and microbial reactions, *Environmental Science Technology* **4**: 929–935.

Des Marais, D. J., Strauss, H., Summons, R. E., and Hayes, J. M., 1992, Carbon isotope evidence for the stepwise oxidation of the Proterozoic environment, *Nature* **359**: 605–608.

Donoghue, P. C. J., Bengtson, S., Dong, X., Gostling, N. J., Huldtgren, T., Cunningham, J. A., Yin, S., Yue, Z., Peng, F., and Stampanoni, M., 2006, Synchrotron X-ray tomographic microscopy of fossil embryos, *Nature* **442**: 680–683.

Durand, B. (Ed.), 1980, *Kerogen – Insoluble Organic Matter from Sedimentary Rocks*, Edition Technip, Paris, 519 p.

Farquhar, J., Bao, H., and Thiemans, M., 2000, Atmospheric influence of Earth's earliest sulfur cycle, *Science* **289**: 756–758.

Fischer, W. W., Summons, R. E., and Pearson, A., 2005, Targeted genomic detection of biosynthetic pathways: anaerobic production of hopanoid biomarkers by a common sedimentary microbe, *Geobiology* **3**: 33–40.

Gibson Jr., E. K., Clemett, S. J., Thomas-Keprta, K. L., McKay, D. S., Wentworth, S. J., Robert, F., Verchovsky, A. B., Wright, I. P., Pillinger, C. T., Rice, T., and Van Loer, B., 2006, Observation and analysis of in situ carbonaceous matter in Naklha: part II, *Luner and Planetary Science* **XXXVII**: A2039.

Goericke, R., Montoya, J. P., and Fry, B., 1994, Physiology of isotopic fractionation in algae and cyanobacteria. In: Lajtha, K., and Michener, R. H. (Eds.) *Stable Isotopes in Ecology and Environmental Science*, Blackwell, Oxford, pp. 187–221.

Grassineau, N. V., Nisbet, E. G., Fowler, C. M. R., Bickle, M. J., Lowry, D., Chapman, H. J., Mattey, D. P., Abell, P., Yong, J., and Martin, A., 2002, Stable isotopes in the Archaean Belingwe Belt, Zimbabwe: evidence for a diverse microbial mat ecology. In: Fowler, C. M. R., Ebinger, C. J., and Hawkesworth, C. J. (Eds.) *The Early Earth: Physical, Chemical and Biological Development, Special Publication of the Geological Society of London* **199**: 309–328.

Habicht, K. S., and Canfield, D. E., 1996, Sulphur isotope fractionation in modern microbial mats and the evolution of the sulphur cycle, *Nature* **382**: 342–343.

Habicht, K. S., and Canfield, D. E., 2001, Isotope fractionation by sulfate-reducing natural populations and the isotopic composition of sulfide in marine sediments, *Geology* **29**: 555–558.

Habicht, K. S., Gade, M., Thamdrup, B., Berg, P., and Canfield, D. E., 2002, Calibration of sulfate levels in the Archean oceans, *Science* **298**: 2373–2374.

Hoefs, J., 1987, *Stable Isotope Geochemistry* 2nd edn., Springer, Berlin, 241p.

Holland, H. D., 2006, The oxygenation of the atmosphere and oceans, *Philosophical Transactions of the Royal Society B* **361**: 903–915.

Horita, J., and Berndt, M. E., 1999, Abiogenic methane formation and isotopic fractionation under hydrothermal conditions, *Science* **285**: 1055–1057.

Jehlicka, J., Urban, O., and Pokorny, J., 2003, Raman spectroscopy of carbon and solid bitumens in sedimentary and metamorphic rocks, *Spectrochimica Acta A* **59**: 2341–2352.

Jia, Y., 2006, Nitrogen isotope fractionations during progressive metamorphism: a case study from the Paleozoic Cooma metasedimentary complex, southeastern Australia, *Geochimica et Cosmochimica Acta* **70**: 5201–5214.

Johnson, C. M., and Beard, B. L., 2005, Biogeochemical cycling of iron isotopes, *Science* **309**: 1025–1027.

Johnson, C. M., Beard, B. L., Klein, C., Beukes, N., and Roden, E., 2007, Iron isotopes constrain the roles of biologic and abiologic processes in formation of banded iron formations, *Geochimica et Cosmoschimica Acta* **71**: A448.

Kasting, J. F., and Ono, S., 2006, Paleoclimates: the first two billion years, *Philosophical Transactions of the Royal Society B* **361**: 917–929.

Kempe, A., Schopf, J. W., Altermann, W., Kudryavstev, A. B., Heckl, W. M., 2002, Atomic force microscopy of Precambrian microscopic fossils, *PNAS* **99**: 9117–9120.

Kerrich, R., Jia, Y., Manikyamba, C., and Naqvi, S. M., 2006, Secular variations of N-isotopes in terrestrial reservoirs and ore deposits. In: Kesler, S. E., and Ohmoto, H. (Eds.) *Evolution of Early Earth's Atmosphere, Hydrosphere, and Biosphere- Constraints from Ore Deposits, GSA Memoir* **198**: 81–104.

Kerridge, J. F., 1985, Carbon, hydrogen and nitrogen in carbonaceous chondrites: abundances and isotopic compositions in bulk samples, *Geochimica et Cosmochimica Acta* **49**: 1707–1714.

Knoll, A. H., 2003, *Life on a Young Planet: The First Three Billion Years of Evolution on Earth*, Princeton University Press, Princeton, NJ, 277 p.

Krot, A. N.,, Keil, K., Goodrich, C. A., Scott, E. R. D., and Weisberg, M. K., 2003, Classification of meteorites. In: Davies, A. M. (Ed.) *Meteorites, Comets and Planets, Treatise on Geochemistry* **1**: 83–128.

Kudryavtsev, A. B., Schopf, J. W., Agresti, D. G., and Wdowiak, T. J., 2001, In situ laser-Raman imagery of Precambrian microscopic fossils, *PNAS* **98**: 823–826.

Lindsay, J. F., Brasier, M. D., McLoughlin, N., Green, O. R., Fogel, M., Steele, A., and Mertzman, S. A., 2005, The problem of deep carbon – an Archaean paradox, *Precambrian Research* **143**: 1–22.

Mango, F. D., 1991, The stability of hydrocarbons under the time-temperature conditions of petroleum genesis, *Nature* **352**: 146–148.

McCollom, T. M., and Seewald, J. S., 2006, Carbon isotope composition of organic compounds produced by abiotic synthesis under hydrothermal conditions, *Earth and Planetary Science Letters* **243**: 64–84.

McKeegan, K. D., Kudryavstev, A. B., and Schopf, J. W., 2007, Raman and ion microscopic imagery of graphitic inclusions in apatite from older than 3830 Ma Akilia supracrustal rocks, west Greenland, *Geology* **35**: 591–594.

Morse, J., and MacKenzie, F., 2002, *Geochemistry of Sedimentary Carbonates*, Elsevier, New York, 707p.

Nemchin, A. A., Whitehouse, M. J., Menneken, M., Geisler, T., Pidgeon, R. T., and Wilde, S. A., 2008, A light carbon reservoir recorded in zircon-hosted diamond from the Jack Hills, *Nature* **454**: 92–95.

Oehler, D. Z., Robert, F., Mostefaoui, S., Meibom, A., Selo, M., and McKay, D. S., 2006, Chemical mapping of Proterozoic organic matter at submicron spatial resolution, *Astrobiology* **6**: 838–850.

Ohmoto, H., and Felder, R. P., 1987, Bacterial activity in the warmer, sulphate-bearing, Archaean oceans, *Nature* **328**: 244–246.

Papineau, D., Mojzsis, S. J., Karhu, J. A., and Marty, B., 2005, Nitrogen isotope composition of ammoniated phyllosilicates: case studies from Precambrian metamorphosed sedimentary rocks, *Chemical Geology* **216**: 37–58.

Pasteris, J. D., and Wopenka, B., 2003, Necessary, but not sufficient: Raman identification of disordered carbon as a signature of ancient life, *Astrobiology* **3**: 727–738.

Pavlov, A. A., Kasting, J. F., Eigenbrode, J. L., and Freeman K. H., 2001, Organic haze in the early atmosphere: sources of low $^{13}C$ late Archean kerogens? *Geology* **29**: 1003–1006.

Philippot, P., Van Zuilen, M., Thomazo, C., Farquhar, J., and Van Kranendonk, M. J., 2007, Early Archaean microorganisms preferred elemental sulfur, not sulfate, *Science* **317**: 1534–1537.

Pinti, D. L., Hashizume, K., and Matsuda, J., 2001, Nitrogen and argon signatures in 3.8 to 2.8 Ga metasediments: clues on the chemical state of the Archean ocean and the deep biosphere, *Geochimica et Cosmochimica Acta* **65**: 2301–2315.

Pinti, D. L., Hashizume, K., Orberger, B., Gallien, J-P., Cloquet, C., and Massault, M., 2007, Biogenic nitrogen and carbon in Fe-Mn-oxyhydroxides from an Archean chert, Marble Bar, Western Australia, *Geochemistry Geophysics Geosystems* **8**: Q02007, doi: 10. 1029/2007/2006GC001394.

Rau, G. H., 1981, Low $^{15}N$:$^{14}N$ in hydrothermal vent animals: ecological implications, *Nature* **289**: 484–485.

Robert, F., Selo, M., Hillion, F., and Skrzypczak, A., 2005, NanoSIMS images of Precambrian fossil cells, *Lunar and Planetary Science* **XXXVI**: A1314.

Rouxel, O. J., Bekker, A., and Edwards, K. J., 2005, Iron isotope constraints on the Archean and Paleoproterozoic ocean redox state, *Science* **307**: 1088–1091.

Rouzaud, J-N., Skrzypczak, A., Bonal, L., Derenne, S., Quirico, E., and Robert, F., 2005, The high resolution transmission electron microscope: a powerful tool for studying the organization of terrestrial and extra-terrestrial carbons, *Lunar and Planetary Science* **XXXVI**: A1322.

Ruchniki, M. D., Elderfield, H., and Spiro, B., 2001, Fractionation of sulfur isotopes during bacterial sulfate reduction in deep ocean sediments at elevated temperatures, *Geochimica et Cosmochimica Acta* **65**: 777–789.

Runnegar B., Dollase, W. A., Ketcham, R. A., Colbert, M., and Carlson, W. D., 2001, Early Archaean sulphates from Western Australia first formed as hydrothermal barites not gypsum evaporites, *Geological Society of America Abstracts* **33**: A-404.

Schidlowski, M., 2001, Carbon isotopes as biogeochemical recorders of life over 3.8 Ga of Earth history: evolution of a concept, *Precambrian Research* **106**: 117–134.

Schopf, J. W., 1993, Microfossils of the Early Archaean Apex Chert: new evidence for the antiquity of life, *Science* **260**: 640–646.

Schopf, J. W., and Kudryavtsev, A. B., 2005, Three-dimensional Raman imagery of Precambrian microscopic organisms, *Geobiology* **3**: 1–12.

Schopf, J. W., Kudryavtsev, A. B., Agresti, D. G., Czaja, A. D., and Wdowiak, T. J., 2005, Raman imagery: a new approach to assess the geochemical maturity and biogenicity of permineralized Precambrian fossils, *Astrobiology* **5**: 333–371.

Schopf, J. W., Tripathi, A. B., and Kudryavtsev, A. B., 2006, Three-dimensional confocal optical imagery of Precambrian microscopic organisms, *Astrobiology* **6**: 1–16.

Schopf, J. W., Kudryavtsev, A. B., Czaja, A. D., and Tripathi, A. B., 2007, Evidence of Archean life: stromatolites and microfossils, *Precambrian Research* **158**: 141–155.

Sephton, M. A., and Gilmour, I., 2001, Compound-specific isotope analysis of the organic constituents in carbonaceous chondrites, *Mass Spectrometry Reviews* **20**: 111–120.

Shen, Y., Buick, R., and Canfield, D. E., 2001, Isotopic evidence for microbial sulphate reduction in the early Archaean era, *Nature* **410**: 77–81.

Sherwood Lollar, B., Westgate, T. D., Ward, J. A., Slater, G. F., and Lacrampe-Couloume, G., 2002, Abiogenic formation of alkanes in the Earth's crust as a minor source for global hydrocarbon reservoirs, *Nature* **416**: 522–524.

Summons, R. E., Jahnke, L. L., Hope, J. M., and Logan, G. A., 1999, 2-methylhopanoids as biomarkers for cyanobacterial oxygenic photosynthesis, *Nature* **400**: 554–557.

Summons, R. E., Bradley, A. S., Jahnke, L. L., and Waldbauer, J. R., 2006, Steroids, triterpenoids and molecular oxygen, *Philosophical Transactions of the Royal Society B* **361**: 951–968.

Tice, M. M., and Lowe, D. R., 2004, Photosynthetic microbial mats in the 3,416-Myr-old ocean, *Nature* **431**: 549–552.

Tice, M. M, Bostick, B. C., and Lowe, D. R., 2004, Thermal history of the 3.5–3.2 Ga Onverwacht and Fig Tree Groups, Barberton greenstone belt, South Africa, inferred by Raman microspectroscopy of carbonaceous material, *Geology* **32**: 37–40.

Tuinstra, F., and Koenig, J. L., 1970, Raman spectrum of graphite, *Journal of Chemical Physics* **53**: 1126–1130.

Ueno, Y., Isozaki, Y., Yurimoto, H., and Maruyama, S., 2001, Carbon isotopic signatures of individual Archean microfossils(?) from Western Australia, *International Geology Review* **43**: 196–212.

Ueno, Y., Yamada, K., Yoshida, N., Maruyama, S., and Isozaki, Y., 2006, Evidence from fluid inclusions for microbial methanogenesis in the early Archaean era, *Nature* **440**: 516–519.

van Zuilen, M. A., Lepland, A., and Arrhenius, G., 2002, Reassessing the evidence for the earliest traces of life, *Nature* **418**: 627–630.

van Zuilen, M. A., Lepland, A., Teranes, J., Finarelli, J., Wahlen, M., and Arrhenius, G., 2003, Graphite and carbonates in the 3.8 Ga old Isua Supracrustal Belt, southern West Greenland, *Precambrian Research* **126**: 331–348.

van Zuilen, M. A., Mathew, K., Wopenka, B., Lepland, A., Marti, K., and Arrhenius, G., 2005, Nitrogen and argon isotopic signatures in graphite from the 3.8-Ga-old Isua Supracrustal Belt, Southern West Greenland, *Geochimica et Cosmochimica Acta* **69**: 1241–1252.

van Zuilen, M. A., Thomazo, C., Luais, B., and Philippot, P., 2007, Carbon and iron isotopic evidence for photosynthesis in a 3.5 Ga old shallow marine depositional environment, *Geochimica et Cosmochimica Acta* **71**: A1058.

Vargas, M., Kashefi, K., Blunt-Harris, E. L., and Lovley, D. R., 1998, Microbiological evidence for Fe(III) reduction of early earth, *Nature* **395**: 65–67.

Wacey, D., Kilburn, M. R., McLoughlin, N., Parnell, J., Stoakes, C. A., and Brasier, M. D., 2008, Use of NanoSIMS to investigate early life on Earth: ambient inclusion trails in a *c.* 3400 Ma sandstone, *Journal of the Geological Society of London* **165**: 43–53.

Westall, F., and Folk, R. L., 2003, Exogenous carbonaceous microstructures in Early Archaean cherts and BIFs from the Isua Greenstone Belt: implications for the search for life in ancient rocks, *Precambrian Research* **126**: 313–330.

Westall, F., de Witt, M. J., Dann, J., van der Gaast, S., de Ronde, C. E. J., and Gerneke, D., 2001, Early Archean fossil bacteria and biofilms in hydrothermally-influenced sediments from the Barberton greenstone belt, South Africa, *Precambrian Research* **106**: 93–116.

Whitehouse, M. J., and Fedo, C. M., 2007, Microscale heterogeneity of Fe isotopes in > 3.71 Ga banded iron formation from the Isua Greenstone Belt, southwest Greenland, *Geology* **35**: 719–722.

Williams, L. B., Ferrel, R. E., Hutcheon, I., Bakel, A. J., Walsh, M. M., and Krouse, H. R., 1995, Nitrogen isotope geochemistry of organic matter and minerals during diagenesis and hydrocarbon migration, *Geochimica et Cosmochimica Acta* **54**: 765–799.

Wopenka, B., and Pasteris, J. D., 1993, Structural characterisation of kerogens to granulite-facies graphite: applicability of Raman microprobe spectroscopy, *American Mineralogist* **78**: 533–557.

Yamaguchi, K. E., Johnson, C. M., Beard, B. L., and Ohmoto, H., 2005, Biogeochemical cycling of iron in the Archean–Paleoproterozoic Earth: constraints from iron isotope variations in sedimentary rocks from the Kaapvaal and Pilbara Cratons, *Chemical Geology* **218**: 135–169.

Yamaguchi, K. E., Johnson, C. M., Beard, B. L., Paulson, S. R., and Ohmoto, H., 2007, Coupled C-S-Fe isotope variations in Archean-Paleoproterozoic shales trace microbial metabolisms and redox state in the early Earth, *Geochimica et Cosmoschimica Acta* **71**: A1136.

Young, J. P. W., 1992, Phylogenetic classification of nitrogen-fixing organisms. In: Stacey, G., Burris, R. H., and Evans, H. J. (Eds.) *Biological Nitrogen Fixation*, Chapman & Hall, New York, pp. 43–86.

# Part II
# An Atlas of Claims for Early Archean Life

This part of the book gives a comprehensive pictorial review of occurrences of putative life in Archean rocks. The images are organised by age (oldest first), according to the rock unit in which they are found. Examples are drawn from Isua and Akilia in Greenland, Pilbara in Australia, and Barberton in South Africa. Younger analogues for the Archean structures, where appropriate, are drawn from all periods of Earth history and are illustrated immediately following the Archean examples. In addition, Chapter 21 contains multiple images of naturally occurring and laboratory constructed non-biological artefacts that may be confused with life.

Each early Archean image is accompanied by a summary table describing the locality and lithology in which the structures are found, their morphology, and any additional techniques (e.g., isotopes) used to support their biogenicity or antiquity. Each table summarises observations available in the literature, and extensive references for further reading are given at the end of each chapter. For the rock units that I have personal experience of and where there have been multiple claims for life, I have provided an extended text summary in addition to the tables and images.

The last line of each summary table is a personal opinion on the reliability of each of the claims for early Archean life based on the extent to which the evidence so far available adheres to the antiquity and biogenicity criteria given in Part I of this book. This is coded for ease of recognition, in different hues of grey, from light grey (not life), through to dark grey (probably life):

| |
|---|
| This is not an example of life before 3,000 Ma. The structure is clearly non-biological and/or a later contaminant |
| This is unlikely to be an example of life before 3,000 Ma. Some evidence may be consistent with early Archean life, but an explanation as a non-biological artefact or a younger contaminant is more plausible |
| This may be an example of life before 3,000 Ma. Such structures exhibit at least one line of evidence for antiquity and biogenicity but more evidence is needed to fully exclude formation as non-biological artefacts or contaminants |
| This is a strong candidate for life before 3,000 Ma. Such structures have clear age constraints and no known non-biological formation mechanism |

It must be emphasised that this colour coding is a personal opinion and is merely intended to make students, and indeed all workers in this field, think critically and cautiously about what they are looking at.

# Chapter 6
# > 3,700 Ma Isua Supracrustal Belt and Akilia Island, S.W. Greenland

**Fig. B1** Putative signs of life have been reported from Akilia Island and the Isua supracrustal belt in south-west Greenland. On Akilia, the claims come from a banded quartz-pyroxene rock that has been interpreted as a metamorphosed banded iron formation. In Isua, claims have come from meta-cherts and meta-BIF together with other banded rocks interpreted as a turbidite sequence (pictured *above*). Controversy surrounds most of these claims with regard to the age, original protolith and biogenicity. Scale: *top*, person is 2 m tall; *bottom*, antlers are 80 cm long (Photographs courtesy of Stephen Moorbath)

D. Wacey, *Early Life on Earth: A Practical Guide*,

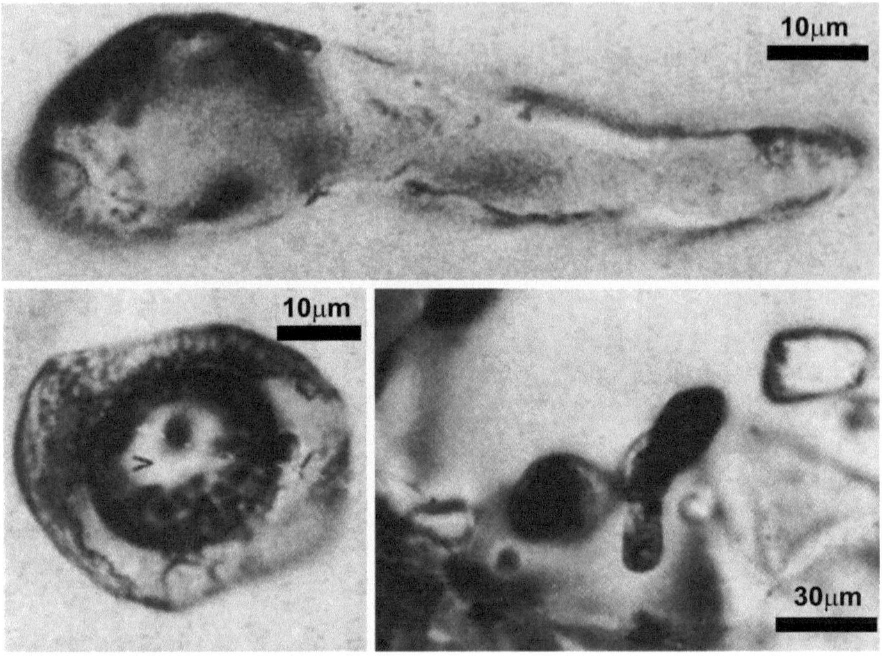


**Fig. B2** Three views of *Isuasphaera* (Photographs reproduced from Pflug and Jaeschke-Boyer, 1979 with permission from Nature Publishing)

| | |
|---|---|
| Age of host rock | Probably >3,700 Ma: Metaquartzite, Isua, Greenland; dated radiometrically but field relations not given |
| Age of structures | Possibly >3,700 Ma: Found below the surface in thin sections; Raman spectra show carbon to be partially graphitised |
| Geological context? | Poorly described – no map or field description; occur in the 'cherty layers of a metaquartzite'; later identified as coming from high strain zone in Isua (Appel et al., 2003a,b) |
| Biological morphology? | Spheres (20–40 μm in diameter) or bulbous filaments (several 10's μm long); sometimes attached in colonies; sometimes contain degraded organic material |
| Biogeochemistry? | The structures are at least in part carbonaceous as indicated by Raman spectra; C–H and C = O bonds also identified by Raman, but these have been attributed to later contamination |
| Type of organism? | Yeast-like (but see below) and named as *Isuasphaera* |
| INTERPRETATION IN THE LITERATURE | 1. Yeast-like fossils (Pflug and Jaeschke-Boyer, 1979)<br>2. Limonite stained fluid inclusions or limonite stained cavities (Bridgewater et al., 1981; Roedder, 1981)<br>3. Post tectonic contaminants (Appel et al., 2003a,b) |
| IS THIS REALLY LIFE? | No, extreme stretching deformation of the host rock cannot possibly have preserved syn-depositional spherical and sub-spherical microfossils |

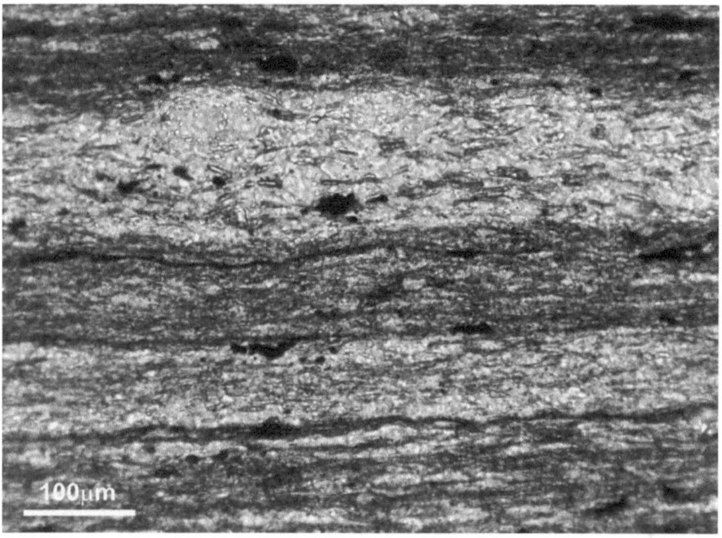


**Fig. B3** Banded 'meta-sediment' with graphite (black) (Sample from the 'Rosing Locality' in Isua, courtesy of Stephen Moorbath)

| | |
|---|---|
| Age of host rock | >3,700 Ma: Garbenschiefer Formation meta-sediments, Isua, Greenland; age constrained by 3,700 Ma U–Pb date on igneous zircons from cross cutting granite |
| Age of structures | Probably > 3,700 Ma: Graphite grains distributed along putative 'sedimentary bands'; graphite also occurs as inclusions in porphyroblasts of biotite and garnet |
| Geological context? | Occur in metamorphosed (amphibolite facies) and deformed remnants of a 50 m succession of volcaniclastic and arguably pelagic sediments; interpreted as a turbidite sequence deposited below wave base; this would be viable for life |
| Biological morphology? | None: Graphite occurs as 2–5 μm sized globules and larger flake-like grains (up to 100 μm) |
| Biogeochemistry? | Carbon isotopes: $^{13}$C averages −19‰ PDB |
| Type of organism? | Carbon derived from unknown source, possibly photoautotrophic organisms (Rosing, 1999) |
| INTERPRETATION IN THE LITERATURE | Biological carbon (Rosing, 1999) although the signature is rather enriched in $^{13}$C |
| IS THIS REALLY LIFE? | Such carbon isotope depletions can be achieved easily without biology in the laboratory; some argument exists over the primary versus secondary nature of the banding in this rock |

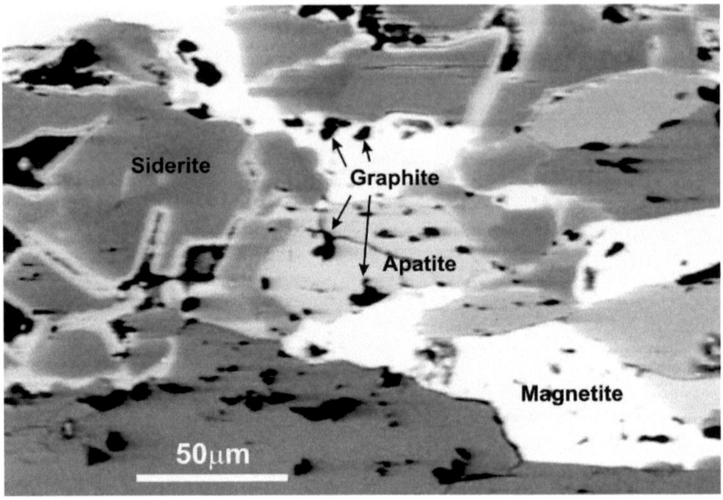


**Fig. B4** Graphite inclusions in apatite and magnetite associated with siderite from Isua (Image modified from Van Zuilen et al., 2002 with permission from Nature Publishing)

| | |
|---|---|
| Age of host rock | >3,700 Ma: 'meta-banded-iron-formation', Isua, Greenland |
| Age of structures | Possibly > 3,700 Ma: Graphite grains occur sealed within apatite (and other phases) but may be affected by later metasomatic fluids |
| Geological context? | Graphite tends to occur in close association with Mg–Mn-siderite and magnetite in meter-wide carbonate-rich layers within volcanic or volcaniclastic rocks. The 'meta-banded-iron-formation' was thus later reclassified as a 'meta-carbonate' |
| Biological morphology? | None: Graphite occurs as tiny inclusions within apatite |
| Biogeochemistry? | Carbon isotopes: $\delta^{13}C$ averages –30‰ ± ‰ PDB, measured *in situ* (Mojzsis et al., 1996)<br>BUT technique has been challenged (see refs. below) and this signal may be due to contamination |
| Type of organism? | Unknown, perhaps not applicable |
| INTERPRETATION IN THE LITERATURE | 1. Biologically derived carbon (and possibly apatite) (Mojzsis et al., 1996; Schidlowski, 1988)<br>2. Carbon formed from thermal-metamorphic disproportionation of ferrous carbonate (siderite) (Van Zuilen et al., 2002, 2003; Lepland et al., 2002) |
| IS THIS REALLY LIFE? | Graphite has only independently been corroborated in carbonate-associated apatites where it likely formed non-biologically; carbon isotope signature may be due to recent contamination |

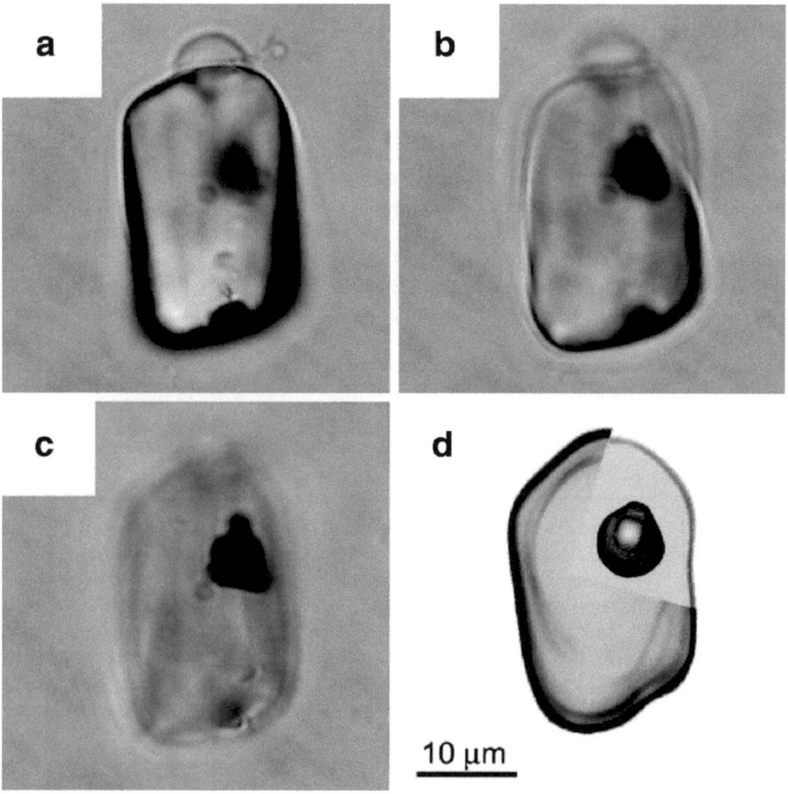


**Fig. B5** Graphite inclusion within ~3,850 Ma apatite grain (Image courtesy of J.W. Schopf from McKeegan et al., 2007)

| | |
|---|---|
| Age of host rock | ~3,850 Ma: Quartz pyroxene rock, Akilia Island, Greenland; age constrained by cross cutting quartz diorite dated at 3,860 ± 10 Ma (although cross cutting relationships on Akilia are controversial) |
| Age of structures | ~3,850 Ma: Graphite occurs within apatite grains included in pyroxene, quartz, amphibole and magnetite |
| Geological context? | This has been heavily debated: Either the rock is a metamorphosed sedimentary banded iron formation, or it is metasomatized ultramafic rock, unsuitable for harbouring traces of life |
| Biological Morphology? | None; the graphite occurs as micron-sized inclusions within lobate apatite shown by the optical (a–c) and confocal Raman (d) images above |
| Biogeochemistry? | Carbon isotopes: Two sets of $\delta^{13}C$ average −37‰ ± 3‰ and −29‰ ± 4‰ PDB respectively, measured *in situ* |
| Type of organism? | Unknown, perhaps not applicable |
| INTERPRETATION IN LITERATURE | 1. Biologically derived carbon (Mojzsis et al., 1996)<br>2. Inorganic carbon (Fedo and Whitehouse, 2002) |
| IS THIS REALLY LIFE? | The geological setting of this unit is very controversial, both in terms of age and primary lithology. In addition, the carbon isotope signature is not uniquely biological |

## Recommended Reading


Appel, P. W. U., Moorbath, S., and Myers, J. S., 2003a, *Isuasphaera isua* (Pflug) revisited, *Precambrian Research* **126**: 309–312.

Appel, P. W. U., Moorbath, S., and Touret, J. L. R., 2003b, Early Archaean processes and the Isua Greenstone Belt, West Greenland, *Precambrian Research* **126**: 173–179.

Bridgewater, D., Allaart, J. H., Schopf, J. W., Klein, C., Walter, M. R., Barghoorn, E. S., Strother, P., Knoll, A. H., and Gorman, B. E., 1981, Microfossil-like objects from the Archaean of Greenland: a cautionary note, *Nature* **289**: 51–53.

Dauphas, N., van Zuilen, M., Wadhwa, M., Davis, A. M., Marty, B., and Janney, P. E., 2004, Clues from Fe isotope variations on the origin of early Archean BIFs from Greenland, *Science* **306**: 2077–2080.

Fedo, C. M., and Whitehouse, M. J., 2002, Metasomatic origin of quartz-pyroxene rock, Akilia, Greenland, and implications for Earth's earliest life, *Science* **296**: 1448–1452.

Lepland, A., Arrhenius, G., and Cornell, D., 2002, Apatite in the Early Archean Isua supracrustal rocks, southern West Greenland: its origin, association with graphite and potential as a biomarker, *Precambrian Research* **118**: 221–241.

Lepland, A., van Zuilen, M. A., Arrhenius, G., Whitehouse, M. J., and Fedo, C. M., 2005, Questioning the evidence for Earth's earliest life—Akilia revisited, *Geology* **33**: 77–79.

Manning, C. E., Mojzsis, S. J., and Harrison, T. M., 2006, Geology, age, and origin of supracrustal rocks at Akilia, West Greenland, *American Journal of Science* **306**: 303–366.

McKeegan, K. D., Kudryavtsev, A. B., and Schopf, J. W., 2007, Raman and ion microscopic imagery of graphitic inclusions in apatite from older than 3830 Ma Akilia supracrustal rocks, west Greenland, *Geology* **35**: 591–594.

Mojzsis, S. J., Arrenhius, G., McKeegan, K. D., Harrison, T. M., Nutman, A. P., and Friend, C. R. L., 1996, Evidence for life on Earth 3,800 million years ago, *Nature* **384**: 55–59.

Moorbath, S., 2005, Dating earliest life, *Nature* **434**: 155.

Moorbath, S., O'Nions, R. K., and Pankhurst, R. J., 1973, Early Archaean age for the Isua iron formation, West Greenland, *Nature* **245**: 138–139.

Myers, J. S., and Crowley, J. L., 2000, Vestiges of life in the oldest Greenland rocks? A review of early Archean geology in the Godthåbsfjord region, and reappraisal of life evidence for > 3850 Ma life on Akilia, *Precambrian Research* **103**: 101–124.

Nutman, A. P., McGregor, V. R., Friend, C. R. L., Bennett, V. C., and Kinny, P. D., 1996, The Itsaq Gneiss Complex of southern west Greenland; the world's most extensive record of early crustal evolution (3900–3600 Ma), *Precambrian Research* **78**: 1–39.

Nutman, A. P., Mojzsis, S. J., and Friend, C. R. L., 1997, Recognition of ≥3850 Ma water-lain sediments in West Greenland and their significance for the Early Archean Earth, *Geochimica et Cosmochimica Acta* **61**: 2475–2484.

Pflug, H. D., and Jaeschke-Boyer, H., 1979, Combined structural and chemical analysis of 3,800-Myr-old microfossils, *Nature* **280**: 483–486.

Roedder, E., 1981, Are the 3,800-Myr-old Isua objects microfossils, limonite-stained fluid inclusions, or neither? *Nature* **293**: 459–462.

Rosing, M. T., 1999, $^{13}$C Depleted carbon microparticles in > 3700-Ma sea-floor sedimentary rocks from West Greenland, *Science* **283**: 674–676.

Rosing, M. T., and Frei, R., 2004, U-rich Archean sea-floor sediments from Greenland – indications of > 3700 Ma oxygenic photosynthesis, *Earth and Planetary Science Letters* **217**: 237–244.

Schidlowski, M., 1988, A 3,800-million-year isotopic record of life from carbon in sedimentary rocks, *Nature* **333**: 313–318.

Ueno, Y., Yurimoto, H., Yoshioka, H., Komiya, T., and Maruyama, S., 2002, Ion microprobe analysis of graphite from ca. 3.8 Ga metasediments, Isua supracrustal belt, West Greenland: relationship between metamorphism and carbon isotopic composition, *Geochimica et Cosmochimica Acta* **66**: 1257–1268.

Van Zuilen, M. A., Lepland, A., and Arhenius, G., 2002, Reassessing the evidence for the earliest traces of life, *Nature* **418**: 627–630.

Van Zuilen, M. A., Lepland, A., Teranes, J., Finarelli, J., Wahlen, M., and Arrhenius, G., 2003, Graphite and carbonates in the 3.8 Ga old Isua Supracrustal Belt, southern West Greenland, *Precambrian Research* **126**: 331–348.

Whitehouse, M. J., and Fedo, C. M., 2003, Deformation features and critical field relationships of Early Archean rocks, southwest Greenland, *Precambrian Research* **126**: 259–271.

# Chapter 7
# ~3,490 Ma Dresser Formation, East Pilbara, Western Australia

**Fig. B6** The Dresser Formation is spatially restricted, only occurring in the North Pole Dome of the East Pilbara. It comprises up to four bedded chert-sulphate-carbonate-jasper units (*lower* image) which can be up to ~30 m thick. These are inter-bedded with pillow basalt and dolerite and intruded and cross cut by multiple generations of hydrothermal feeder veins. Putative signs of life have been reported from the lowermost chert-barite unit (*upper* image) and associated feeder veins. Scale: camera lens cover is 7 cm (Photographs courtesy of Martin van Kranendonk and Matt Kilburn)

D. Wacey, *Early Life on Earth: A Practical Guide*,

## 7.1 Summary of Claims for Early Life from this Formation

Nodular and wavy-laminated stromatolites (Fig. B8) were first reported from a chert unit within the Dresser Formation at North Pole by Walter et al. (1980). A biological explanation was put forward based upon the close macro-morphological similarity with younger unambiguous biological stromatolites. However, when subjected to re-examination, doubts were raised regarding their biogenicity. For example, Lowe (1994) re-interpreted the Dresser stromatolites as produced by soft sediment deformation of originally flat layers.

Van Kranendonk (2006) showed that a number of Dresser Formation stromatolites (Fig. B9) occur in the vents of barite dykes and suggested that these may have been constructed by hyper-thermophilic microbes. The biogenicity of these stromatolites is again uncertain, because their macro-morphology appears to be largely controlled by the thickness of precipitated barite crusts and draping chert layers. Their distribution could simply reflect the supply of supersaturated solutions from which they precipitated. Robust micro-textural and isotopic evidence for the involvement of any kinds of microbes in the growth of these stromatolites is still lacking.

In terms of microfossils, minute spheroids have been described from the Dresser Formation (Fig. B16; Dunlop et al., 1978) whose size distribution and kerogenous composition were used to infer biogenicity. However, both the syngenicity and biogenicity of these structures were later questioned and they were re-interpreted as either viscous bitumen droplets in secondary megaquartz and chalcedony laminae (Buick, 1990), or simple mineralic non-biological spheroids (Awramik et al., 1983).

A number of structures interpreted as microfossil-like artefacts have been described from this Formation, notably by Buick (1990). These include solid 'inorganic' carbon filaments (Fig. B17) and crystal rim artefacts around spherulitic silica (Fig. B14). These highlight the immense difficulty in identifying microfossils based on morphology alone. Buick's work also highlights the importance of accurate and detailed geological mapping, especially in this most complicated of areas. Nonetheless, other occurrences of microfossils have been claimed (Figs. B12 and B13; Ueno et al., 2001a, b) which await independent verification. Ambient inclusion trails (AITs) have also been illustrated from this unit (Fig. B15; Awramik et al., 1983; Buick, 1990). As mentioned previously (Section 1.3 and see Figs. B45 and B46), recent work (Wacey et al., 2008) suggests that there may be a biological component to these enigmatic structures, but more work is needed at this locality to constrain both their biogenicity and antiquity.

Potential biominerals have recently been reported from one of the barite beds within the Dresser Formation. Sulphur isotope fractionations ($\delta^{34}S$) of up to around 22‰ between co-occurring pyrite and barite (Fig. B11) have been used to argue for the presence of sulphate reducing bacteria at this time (Shen et al., 2001; see Section 5.14 for the theory behind this claim). As an extension to this work, minor sulphur isotopes ($\delta^{33}S$ and $\delta^{36}S$) have been used to argue for microbial sulphur disproportionation pathways in addition to (or possibly in place of) biological sulphate reduction (Philippot et al., 2007). Some argument exists, however, around the interpretation of the barite being a hydrothermal replacement of an original low temperature gypsum mineralogy (Shen et al., 2001; Runnegar et al., 2001), which is a key observation to discount higher temperature non-biological sulphur isotope fractionation.

The latest claim for life from this unit uses the very negative carbon isotope signature ($\delta^{13}C$ as low as −56‰) of methane trapped within fluid inclusions in a silica vein (Fig. B7) to argue for the presence of methanogenic microbes at 3,490 Ma (Fig. B18; Ueno et al., 2006; see also Section 5.13). This is a well reasoned claim with strong evidence for the antiquity of the methane. Sherwood Lollar and McCollum (2006) raise a valid point, however, stating that similar isotopic values have been found both in the field and in the laboratory for methane derived non-biologically. Although Ueno et al. counter with field evidence against one particular non-biological reaction, it is yet to be seen whether the doubters have been satisfied.

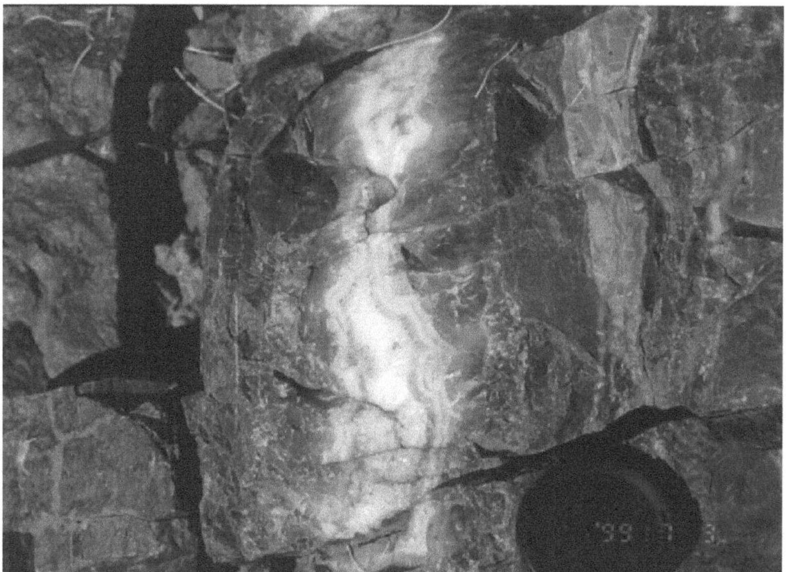

**Fig. B7** A typical hydrothermal silica vein from the Dresser Formation. The majority of the vein is composed of very fine grained, black, carbonaceous chert. The central portion (white area) is made up of agate and coarse grained quartz. Fluid inclusions containing putative microbial methane (Ueno et al., 2006) occur in this central portion (Photograph courtesy of Yuichiro Ueno)

**Fig. B8** Putative stromatolite from North Pole, Pilbara (Photograph courtesy of Matt Kilburn)

| | |
|---|---|
| Age of host rock? | ~3,490 Ma: Dresser Formation, North Pole, Pilbara, Western Australia; age from lead (Pb) isotopes in galena |
| Age of structures? | ~3,490 Ma: Primary sedimentary structure; adjacent and overlying beds show no deformation |
| Geological context? | Interpreted to have been deposited in a shallow-water to intermittently exposed environment, although intimately associated with hydrothermal vent deposits; any microbes likely needed to be thermophilic |
| Biological morphology? | Macro-morphology varies between domical, nodular and stratiform; microstructure consists of wavy and wrinkled laminae, mostly 50–200 μm thick, plus domical structures up to 3 cm wide that persist vertically to produce pseudo-columns; intraclasts found alongside have near identical microstructure; closely comparable in morphology to younger biological stromatolites |
| Biogeochemistry? | None |
| Type of organism? | Unknown but comparisons made to structures built by the filamentous bacterium *Chloroflexus* |
| INTERPRETATION IN THE LITERATURE | 1. Biological stromatolite (e.g., Walter et al., 1980)<br>2. Non-biological stromatolite (e.g., Lowe, 1994) |
| IS THIS REALLY LIFE? | Over-reliance on morphology. Simple stromatolites such as this have been produced without biology (see Chapter 21) |

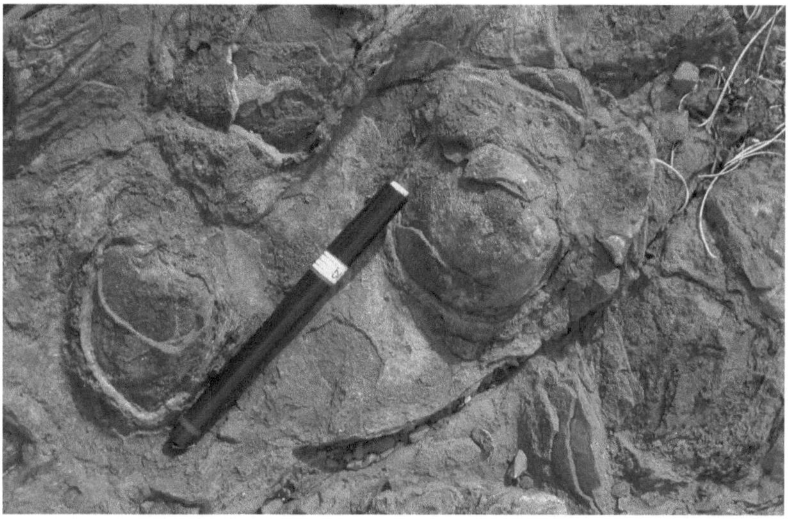

**Fig. B9** Putative conical stromatolites from North Pole, Pilbara (Image courtesy of Martin van Kranendonk)

| | |
|---|---|
| Age of host rock? | ~3,490 Ma: Dresser Formation, North Pole, Pilbara, Western Australia |
| Age of structures? | ~3,490 Ma: Primary sedimentary structure; adjacent and overlying beds show no deformation |
| Geological context? | Chemical precipitation within a restricted basin; intimately associated with hydrothermal vent deposits; any microbes likely needed to be thermophilic |
| Biological morphology? | Steep sided conical; compared by some to younger biological conical stromatolites (e.g., *Conophyton*). |
| Biogeochemistry? | None |
| Type of organism? | Unknown (chemotrophic and thermophilic organism implied from depositional setting) |
| INTERPRETATION IN THE LITERATURE | Biological stromatolite (Van Kranendonk, 2006) |
| IS THIS REALLY LIFE? | Possible life, but stromatolite-like structures can be produced without biology (although cones specifically have yet to be) |

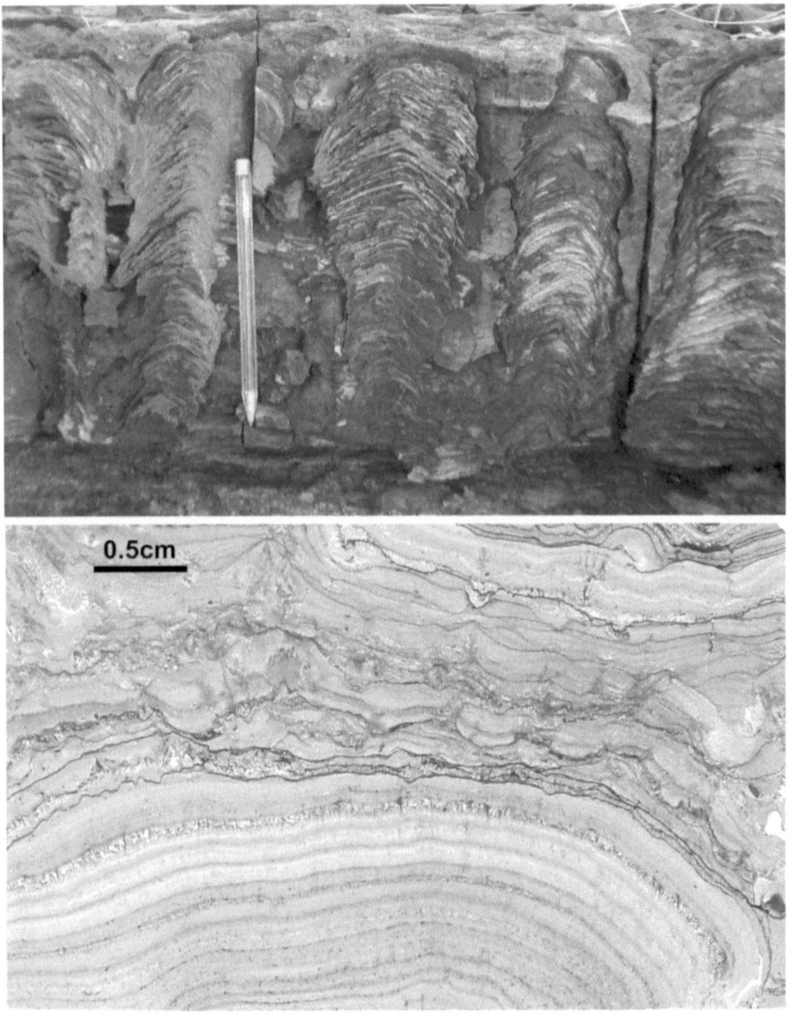


**Fig. B10** Younger partial analogues for the stromatolites described from the Dresser Formation. *Top*, columnar stromatolite from the 2,715 Ma Meentheena carbonate member of the Tumbiana Formation, Fortescue Group, Western Australia. Scale: pen is 12 cm long (Image courtesy of the Western Australia Geological Survey). *Bottom*, thin section of a Meentheena stromatolite from the Oxford Palaeobiology Collections. Note the complex laminae on a variety of scales. Such laminae are not observed in thin sections of the Dresser 'stromatolites', although this may simply be due to poorer preservation

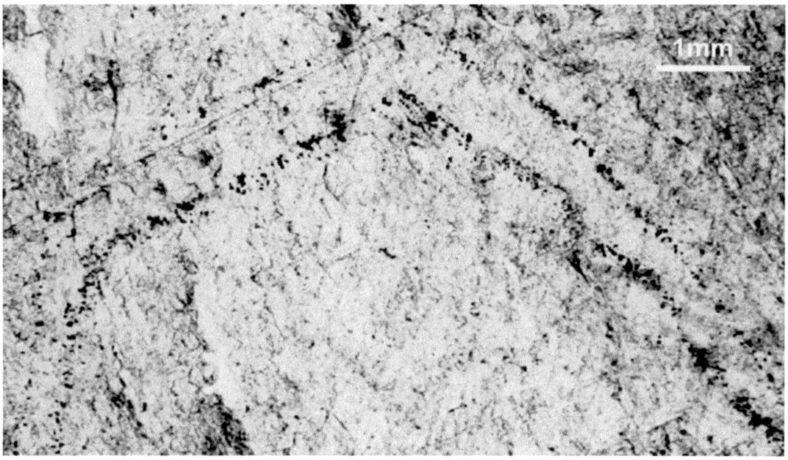


**Fig. B11** Putative pyrite biominerals (black) aligned along sulphate crystal faces from the Dresser Formation (Image courtesy of Roger Buick)

| | |
|---|---|
| Age of host rock? | ~3,490 Ma: Dresser Formation, North Pole, Pilbara, Western Australia |
| Age of structures? | ~3,490 Ma: Microscopic pyrites are aligned along the growth faces of original sulphate crystals in a bedded unit |
| Geological context? | Debated: The pyrite is found within lenses of bedded barite. The barite is interpreted to be a replacement of gypsum that was originally precipitated in an evaporitic brine pond. BUT a primary hydrothermal precipitate mechanism has also been invoked for the barite (Runnegar et al., 2001). |
| Biological morphology? | None |
| Biogeochemistry? | Sulphur isotope fractionation of 21.1–7.4‰ between pyrite (enriched in $^{32}S$) and co-existing sulphate; this is consistent with biological sulphate-reduction; pyrite is associated with organic carbon |
| Type of organism? | Pyrite formed as a consequence of the metabolic activity of sulphate-reducing bacteria |
| INTERPRETATION IN THE LITERATURE | Biominerals (Shen et al., 2001; Shen and Buick, 2004; Philippot et al., 2007) |
| IS THIS REALLY LIFE? | A solid claim for life, although the magnitude of sulphur isotope fractionation is not uniquely biological. A biological interpretation is heavily reliant on the geological context (which is disputed) |

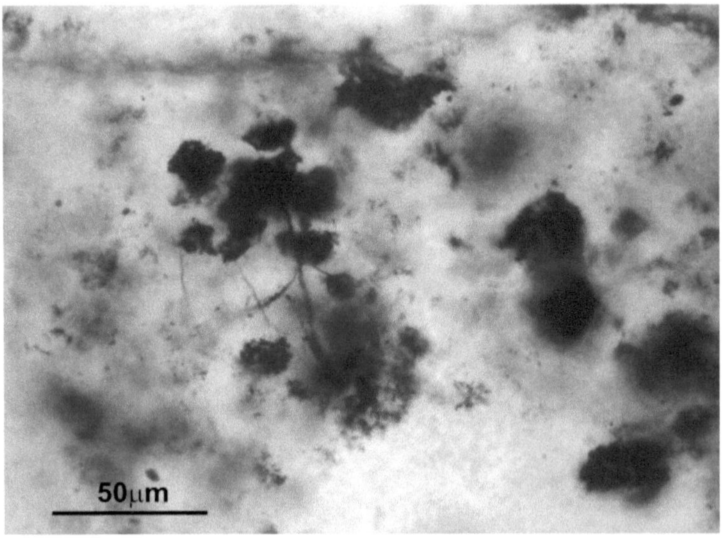


**Fig. B12** Putative filamentous microfossils from the Dresser Formation (Image courtesy of Y. Ueno)

| | |
|---|---|
| Age of host rock? | ~3,490 Ma: Dresser Formation, North Pole, Pilbara, Western Australia |
| Age of structures? | ~3,490 Ma: Found within black silica veins that have intruded inter-bedded chert and barite; cross cutting relationships suggest vein formation was synchronous to the deposition of the bedded chert; although the detailed stratigraphy is poorly understood, a 3,490 Ma age has been obtained from galena within the chert-barite unit |
| Geological context? | Not in a sedimentary unit; the silica veins were hydrothermally emplaced (although this does not necessarily exclude it from being viable for life) |
| Biological morphology? | Filaments: Average 1 μm in diameter and 10–100 μm in length; some have three dimensional spiral geometries; some are mutually interwoven; some are branched; some radiate from kerogen clots (pictured above) |
| Biogeochemistry? | $\delta^{13}C$ = −42‰ to −32‰ PDB, measured *in situ*; consistent with reductive acetyl-CoA and Calvin Cycle autotrophic biological pathways |
| Type of organism? | Comparisons made with several modern filamentous bacteria but simple morphology and preservation preclude identification |
| INTERPRETATION IN THE LITERATURE | Probable microfossils (Ueno et al., 2001a) |
| **IS THIS REALLY LIFE?** | Possible life, although filaments appear solid with no cellular features |

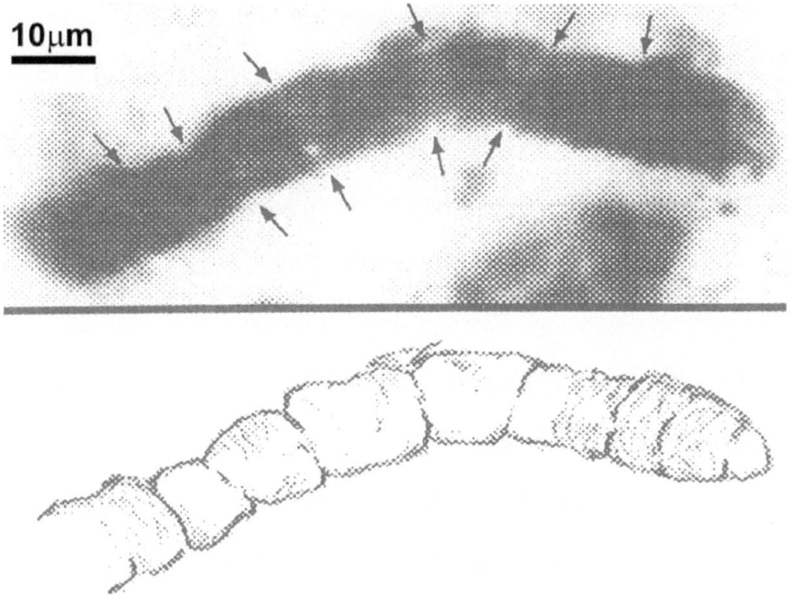


**Fig. B13** Thin section image and interpretive sketch of a filamentous microfossil from the Dresser Formation (Image courtesy of Y. Ueno)

| | |
|---|---|
| Age of host rock? | ~3,490 Ma: Dresser Formation, North Pole, Pilbara, Western Australia |
| Age of structures? | ~3,490 Ma: Found permineralized in chert within an interbedded chert-barite unit, and in chert clasts within an overlying breccia; although the detailed stratigraphy is poorly understood, a 3,490 Ma age has been obtained from galena within the chert-barite unit |
| Geological context? | Bedded chert chemically precipitated from silica rich hydrothermal fluids; these fluids could also provide various electron donors and acceptors to chemoautotrophic microbes; hence, viable for life |
| Biological morphology? | Filaments – Type A (pictured above): Segmented, unbranched, average 8.7 μm in diameter; segments cylindrical with hemispherical terminus<br>– Type B: Non-segmented, tubular, average 12.6 μm in diameter; lengths not reported |
| Biogeochemistry? | $\delta^{13}C = -40.2‰$ to $-30.5‰$ PDB measured on acid-insoluble residues (i.e. not *in situ*). Consistent with reductive acetyl-CoA and Calvin Cycle autotrophic biological pathways |
| Type of organism? | Comparisons made with *beggiatoacean* filamentous bacteria (Ueno et al., 2001b) |
| INTERPRETATION IN THE LITERATURE | Probable microfossils (Ueno et al., 2001b) |
| **IS THIS REALLY LIFE?** | Possible life, but a non-biological formation mechanism (cf. Apex chert microstructures on pages 12–15 and 156–158) is also highly possible |

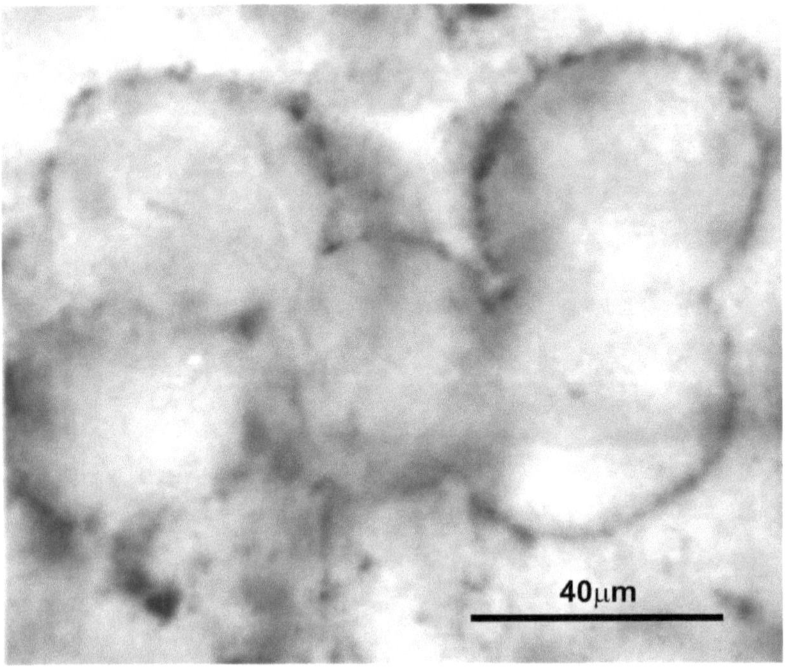


**Fig. B14** Spherical pseudofossils from the Dresser Formation (Image reproduced with permission from Roger Buick and under the fair usage policy of the SEPM Society for Sedimentary Geology)

| | |
|---|---|
| Age of host rock? | ~3,490 Ma: Dresser Formation, Pilbara, Western Australia |
| Age of structures? | May be much younger than 3,490 Ma: occur within secondary (i.e. later) banded chert |
| Geological context? | The kerogenous chert host comes from an overlying fissure. It likely sank down from this fissure into its current position |
| Biological morphology? | Hollow spheroidal: Around 10–50 μm in diameter with a 0.5–1 μm exterior wall; colonial with adjacent spheroids often attached through perforated walls. |
| | BUT: No layered walls, no external sculpture, no internal bodies, no protuberances, no vesicles or partitions |
| Biogeochemistry? | Kerogenous |
| Type of organism? | Comparisons made by some to coccoid microfossils *Clonophycus* and *Eoentophysalis belcherensis* (but see below) |
| INTERPRETATION IN THE LITERATURE | Pseudofossils formed around chalcedonic silica spherulites (Buick, 1990) |
| **IS THIS REALLY LIFE?** | No, both biogenicity and antiquity are unproven |

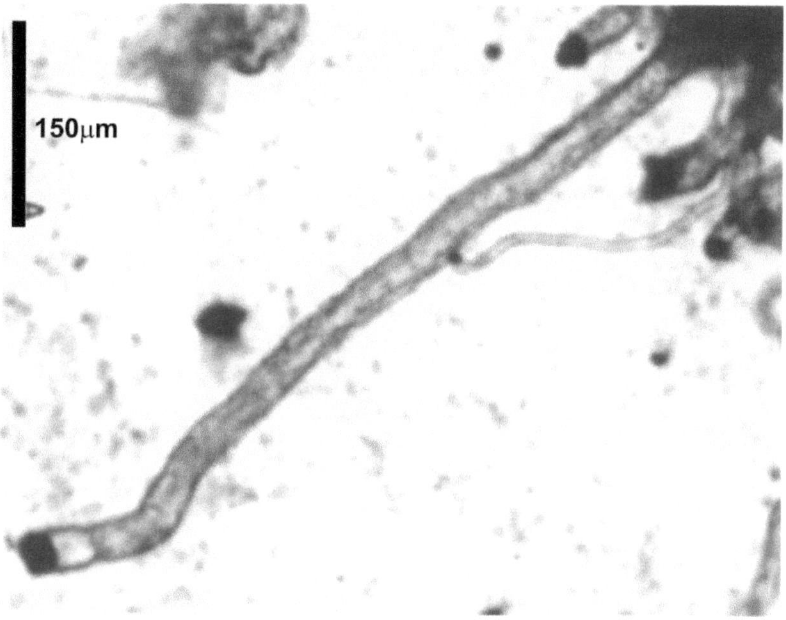


**Fig. B15** Ambient inclusion trails from the Dresser Formation (Image reproduced with permission from Roger Buick and under the fair usage policy of the SEPM Society for Sedimentary Geology)

| | |
|---|---|
| Age of host rock? | ~3,490 Ma: Dresser Formation, Pilbara, Western Australia |
| Age of structures? | Unknown: Occur in a gelatinous precipitate known only from secondary chert veins in the region; probably emplaced some time after the sedimentary chert barite unit was deposited |
| Geological context? | Probably not synchronous with sedimentary unit (see above) |
| Biological morphology? | Broad tubular filaments: 8–30 μm in diameter, 20–800 μm long; have constant diameter and do not branch; can be sinuous, or straight and often kinked; sometimes form radiating clusters from pyrite or kerogen clots; BUT have polygonal cross sections and terminal pyrite crystals, which are not characteristically biological |
| Biogeochemistry? | None – not kerogenous |
| Type of organism? | Not applicable |
| INTERPRETATION IN THE LITERATURE | Ambient inclusion trails (AIT) whose formation mechanism is poorly understood (Awramik et al., 1983; Buick, 1990) |
| **IS THIS REALLY LIFE?** | Both antiquity and biogenicity are unproven at this locality – although see pages 183 and 187 for the possible role of biology in AIT formation |

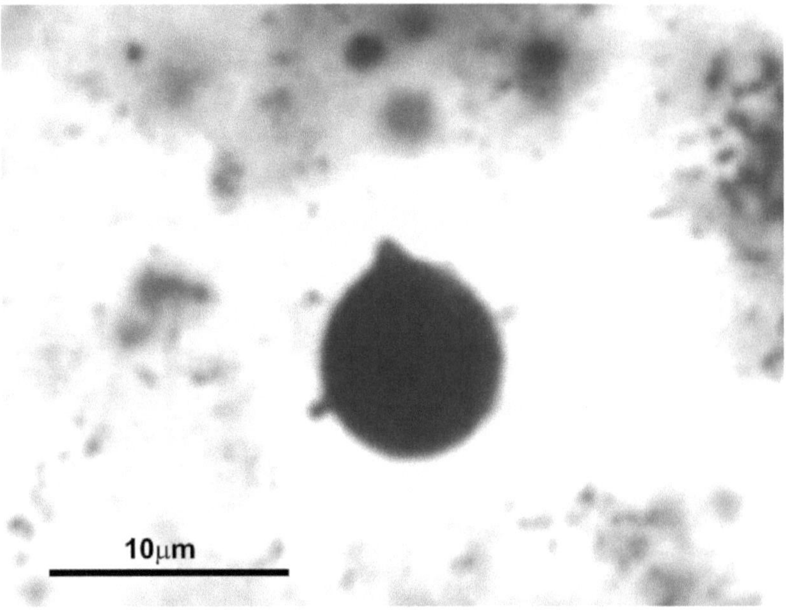


**Fig. B16** A carbonaceous globule from the Dresser Formation (Image reproduced with permission from Roger Buick and under the fair usage policy of the SEPM Society for Sedimentary Geology)

| | |
|---|---|
| Age of host rock? | ~3,490 Ma: Dresser Formation, Pilbara, Western Australia |
| Age of structures? | Probably younger than ~3,490 Ma: Laminae containing the spheroids may not be a primary sediment |
| Geological context? | Occur within a kerogenous chert that has likely been emplaced after sedimentation; may or may not be viable for life |
| Biological morphology? | Spherical, elliptical and reniform: 0.2–7.2 μm in diameter; smooth surfaces except for occasional spinose protuberances; can occur in pairs, chains or clusters. BUT: Many are solid, found at the intersection of grain boundaries and show gradation into irregular kerogen particles |
| Biogeochemistry? | Made of kerogen |
| Type of organism? | Interpreted by some as coccoid bacteria (Dunlop et al., 1978) |
| INTERPRETATION IN THE LITERATURE | 1. Microfossils (Dunlop et al., 1978)<br>2. Non-biological solid carbonaceous globules (Schopf and Walter, 1983)<br>3. Bitumen droplets (Buick, 1990) |
| **IS THIS REALLY LIFE?** | Unlikely, both biogenicity and antiquity are highly questionable |

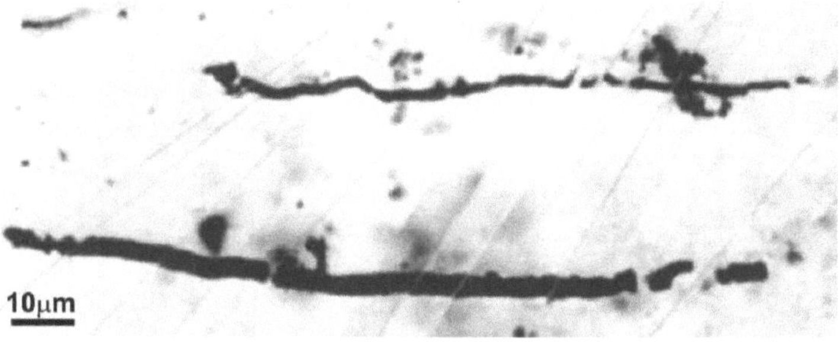


**Fig. B17** Probable inorganic filaments from the Dresser Formation (Image reproduced with permission from Roger Buick and under the fair usage policy of the SEPM Society for Sedimentary Geology)

| | |
|---|---|
| Age of host rock? | ~3,490 Ma: Dresser Formation, Pilbara, Western Australia |
| Age of structures? | May be much younger than 3,490 Ma: Occur within lenses of secondary (i.e. younger) chert |
| Geological context? | The kerogenous chert host appears to have replaced evaporite nodules in a carbonate mudstone; environment may or may not be viable for life |
| Biological morphology? | Filaments of two size ranges:<br>1. Broad (pictured above): 1–5 μm in diameter, 30–70 μm long; straight or sinuous; unbranched; solitary; in segments; circular in cross section<br>2. Fine: <1 μm in diameter, up to 135 μm long; straight or sinuous; unbranched; cylindrical; clustered; disrupted into rod-like segments<br>BUT: Solid, smooth, neither widespread nor numerous |
| Biogeochemistry? | Made of kerogen |
| Type of organism? | Superficial similarity to sheaths of filamentous bacteria |
| INTERPRETATION IN THE LITERATURE | 1. Inorganic filaments (Buick, 1990)<br>2. Possible microfossils (Awramik et al., 1983) |
| IS THIS REALLY LIFE? | Unlikely, great care must be taken to distinguish inorganic filaments from filamentous microbes. Age relationships at this locality are also particularly complicated |

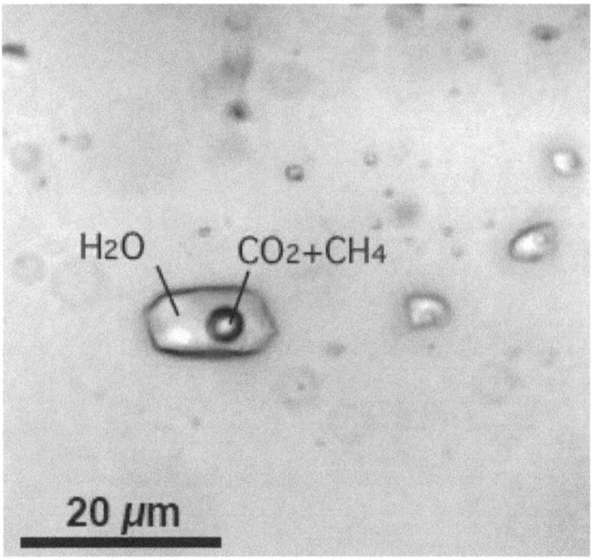


**Fig. B18** Fluid inclusion within quartz containing $CO_2$ and $CH_4$ (Photograph courtesy of Yuichiro Ueno)

| | |
|---|---|
| Age of host rock? | ~3,490 Ma: Dresser Formation, Pilbara, Western Australia; age from lead (Pb) isotopes in galena within the formation |
| Age of structures? | ~3,490 Ma: Fluid inclusions within quartz; found along growth faces and orientated in growth direction of the quartz, hence interpreted as primary |
| Geological context? | Fluid inclusions found in silica dykes interpreted to have been carrying low temperature (<200°C) hydrothermal silica-rich fluids; methane produced from microbes may have become trapped during formation of the dykes |
| Biological morphology? | None |
| Biogeochemistry? | Carbon isotopes from methane ($CH_4$) trapped in fluid inclusions: $\delta^{13}C = -56‰$ to $-36‰$ PDB; consistent with microbial methanogenesis<br>BUT similar values have been produced in non-biological laboratory synthesis |
| Type of organism? | Methanogens (Ueno et al., 2006) |
| INTERPRETATION IN THE LITERATURE | 1. Biological methane trapped within fluid inclusions (Ueno et al., 2006)<br>2. Non-biological methane (Sherwood Lollar and McCollum, 2006) |
| **IS THIS REALLY LIFE?** | Possible life, but both biological and non-biological explanations exist. Use of only one line of evidence (e.g., C-isotopes) often leads to equivocal results |

# Recommended Reading


Awramik, S. M., Schopf, J. W., and Walter, M. R., 1983, Filamentous fossil bacteria from the Archaean of Western Australia, *Precambrian Research* **20**: 357–374.

Buick, R., 1990, Microfossil recognition in Archean rocks: an appraisal of spheroids and filaments from a 3500 M.Y. old chert-barite unit at North Pole, Western Australia, *Palaios* **5**: 441–459.

Dunlop, J. S. R., Muir, M. D., Milne, V. A., and Groves, D. I., 1978, A new microfossil assemblage from the Archaean of Western Australia, *Nature* **274**: 676–678.

Lowe, D. R., 1994, Abiological origin of described stromatolites older than 3.2 Ga, *Geology* **22**: 387–390.

Philippot, P., Van Zuilen, M., Lepot, K., Thomazo, C., Farquhar, J., and Van Kranendonk, M. J., 2007, Early Archaean microorganisms preferred elemental sulfur, not sulfate, *Science* **317**: 1534–1537.

Runnegar, B., Dollase, W. A., Ketcham, R. A., Colbert, M., and Carlson, W. D., 2001, Early Archaean sulphates from Western Australia first formed as hydrothermal barites not gypsum evaporites, *Geological Society of America Abstracts* **33**: A-404.

Schopf, J. W., and Walter, M. R., 1983, Archean microfossils: new evidence of ancient microbes. In: Schopf, J. W. (Ed.) *Earth's Earliest Biosphere, Its Origin and Evolution*, Princeton University Press, Princeton, NJ, pp. 214–239.

Shen, Y., and Buick, R., 2004, The antiquity of microbial sulfate reduction, *Earth Science Reviews* **64**: 243–272.

Shen, Y., Buick, R., and Canfield, D. E., 2001, Isotopic evidence for microbial sulphate reduction in the early Archean era, *Nature* **410**: 77–81.

Sherwood Lollar, B., and McCollum, T. M., 2006, Biosignatures and abiotic constraints on early life, *Nature* **444**: E18.

Ueno, Y., Isozaki, Y., Yurimoto, H., and Maruyama, S., 2001a, Carbon isotopic signatures of individual Archean microfossils (?) from Western Australia, *International Geology Reviews* **43**: 196–212.

Ueno, Y., Maruyama, S., Isozaki, Y., and Yurimoto, H., 2001b, Early Archean (ca.3.5 Ga) microfossils and $^{13}C$-depleted carbonaceous matter in the North Pole area, Western Australia: field occurrence and geochemistry. In: Nakashima, S., Maruyama, S., Brack, A., and Windley, B. F. (Eds.) *Geochemistry and the Origin of Life*, Universal Academy Press, Tokyo, pp. 203–236.

Ueno, Y., Yoshioka, H., Maruyama, S., and Isozaki, Y., 2004, Carbon isotopes and petrography of kerogens in ~3.5 Ga hydrothermal silica dykes in the North Pole area, Western Australia, *Geochimica et Cosmochimica Acta* **68**: 573–589.

Ueno, Y., Yamada, K., Yoshida, N., Maruyama, S., and Isozaki, Y., 2006, Evidence from fluid inclusions for microbial methanogenesis in the early Archaean era, *Nature* **440**: 516–519.

Van Kranendonk, M. J., 2006, Volcanic degassing, hydrothermal circulation and the flourishing of early life on Earth: a review of the evidence from c. 3490–3240 Ma rocks of the Pilbara Supergroup, Pilbara Craton, Western Australia, *Earth Science Reviews* **74**: 197–240.

Wacey, D., Kilburn, M. R., McLoughlin, N., Parnell, J., Stoakes, C. A., and Brasier, M. D., 2008, Use of NanoSIMS to investigate early life on Earth: ambient inclusion trails in a *c.* 3400 Ma sandstone, *Journal of the Geological Society of London* **165**: 43–53.

Walter, M. R., 1983, Archean stromatolites: evidence of the earth's earliest benthos. In: Schopf, J. W. (Ed.) *Earth's Earliest Biosphere, Its Origin and Evolution*, Princeton University Press, Princeton, NJ, pp. 187–213.

Walter, M. R., Buick, R., and Dunlop, J. S. R., 1980, Stromatolites, 3,400–3,500 Myr old from the North Pole area, Western Australia, *Nature* **284**: 443–445.

# Chapter 8
# ~3,470 Ma Mount Ada Basalt, East Pilbara, Western Australia

**Fig. B19** The Mount Ada Basalt is up to 3–4 km thick and typically consists of massive tholeiitic basalt and dolerite intrusions. Occasionally volcanic textures such as vesicles and pillows are preserved. Most of the basalt is extensively recrystallized, folded and carbonate altered. Chert inter-beds are rare but putative evidence for life, in the form of filamentous microfossils and stratiform stromatolites, has been reported from one such banded chert in the North Pole Dome. These are highly controversial because the exact locality has yet to be relocated and recollected. Indeed, the samples may not even be from the Mt. Ada Formation!
Photograph shows dirt track into the North Pole Dome, Pilbara, Western Australia with 4 × 4 vehicle (approximately *centre*) for scale (Photograph by the author)

D. Wacey, *Early Life on Earth: A Practical Guide*,

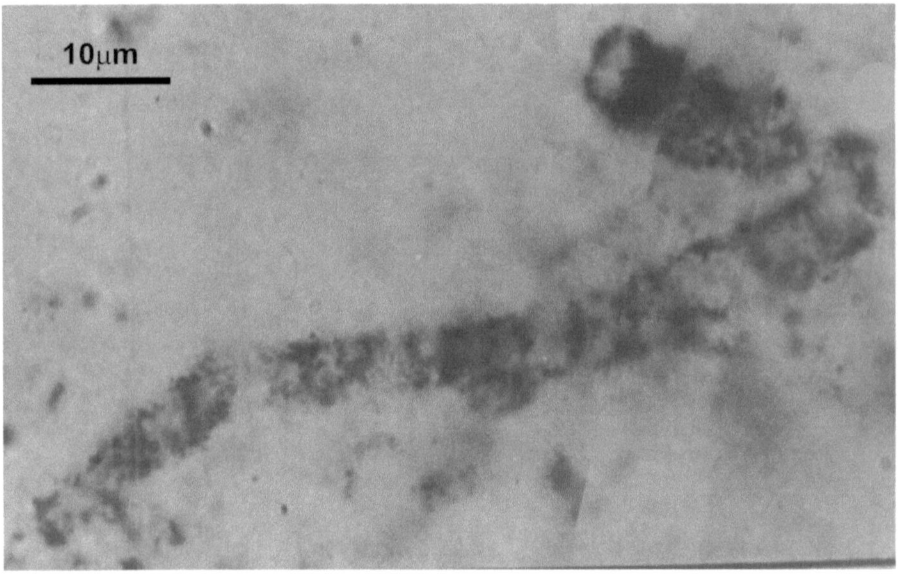


**Fig. B20** Putative microfossil from unknown locality possibly within the Mt. Ada Basalt (Photograph reproduced from Awramik et al., 1983, with permission from Elsevier)

| | |
|---|---|
| Age of host rock? | Unknown: Sample was collected without any detailed geological map or GPS co-ordinates. Several attempts have subsequently failed to recollect the material. It may be from a chert unit from within the ~3,470 Ma Mount Ada Basalt but it may also be from a much younger chert from the late Archean Fortescue Group |
| Age of structures? | Unknown (see above) |
| Geological context? | Unknown (see above), most likely from secondary chert |
| Biological morphology? | Filamentous: Initially reported as numerous, of diverse form and comparable to distinct living taxa; independent re-examination found only one filamentous structure which is arguably septate (pictured above) |
| Biogeochemistry? | Carbonaceous composition |
| Type of organism? | Comparisons made in initial report to both bacteria and cyanobacteria |
| INTERPRETATION IN THE LITERATURE | 1. Microfossil (Awramik et al., 1983)<br>2. Younger contaminant (Buick, 1990) |
| **IS THIS REALLY LIFE?** | The inability to relocate/recollect this sample means that it cannot, as yet, be used as evidence for early Archean life |

**Fig. B21** Paired spheroids, possibly from chert within the Mt. Ada Basalt (Photograph reproduced from Awramik et al., 1983, with permission from Elsevier. Scale bar is 10 μm)

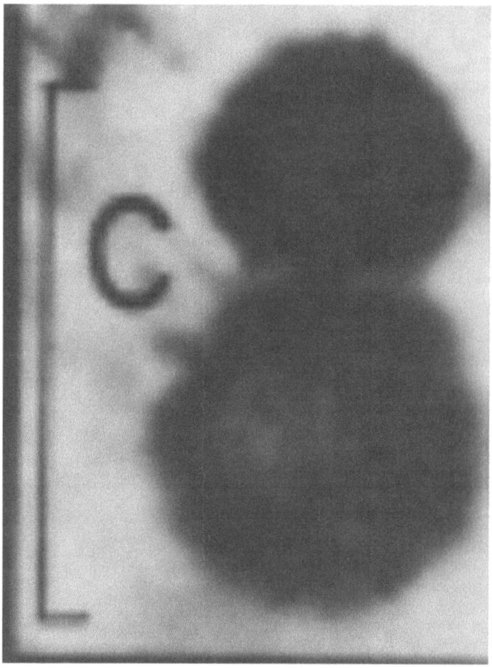


| | |
|---|---|
| Age of host rock? | Unknown: Sample was collected without any detailed geological map or GPS co-ordinates. Several attempts have subsequently failed to recollect the material. It may be from a chert unit from within the ~3,470 Ma Mount Ada Basalt but it may also be from a much younger chert from the late Archean Fortescue Group |
| Age of structures? | Unknown (see above); brownish in colour so likely rather young |
| Geological context? | Unknown (see above) |
| Biological morphology? | Abundant, isolated or paired spheroids; hollow, 4–14 μm in diameter with thin granular wall |
| Biogeochemistry? | Spheroids are made of kerogen |
| Type of organism? | Named as *Archaeosphaeroides pilbarensis* (coccoid bacteria) in initial report by Awramik et al. (1983) |
| INTERPRETATION IN THE LITERATURE | 1. Possible microfossil (Awramik et al., 1983)<br>2. Younger contaminant of unproven biogenicity (Buick, 1990) |
| **IS THIS REALLY LIFE?** | The inability to relocate/recollect this sample means that it cannot, even if biogenicity is proven, be used as evidence for early Archean life |

## Recommended Reading

Awramik, S. M., 1992, The oldest records of photosynthesis, *Photosynthesis Research* **33**: 75–89.

Awramik, S. M., Schopf, J. W., and Walter, M. R., 1983, Filamentous fossil bacteria from the Archaean of Western Australia, *Precambrian Research* **20**: 357–374.

Awramik, S. M., Schopf, J. W., and Walter, M. R., 1988, Carbonaceous filaments from North Pole, West Australia: are they fossil bacteria in Archean stromatolites? A discussion, *Precambrian Research* **39**: 303–309.

Buick, R., 1984, Carbonaceous filaments from North Pole, Western Australia: are they fossil bacteria in Archaean stromatolites? *Precambrian Research* **24**: 157–172.

Buick, R., 1988, Carbonaceous filaments from North Pole, West Australia: are they fossil bacteria in Archean stromatolites? A reply, *Precambrian Research* **39**: 311–317.

Buick, R., 1990, Microfossil recognition in Archaean rocks: an appraisal of spheroids and filaments from a 3500 MY old chert-barite unit at North Pole, Western Australia, *Palaios* **5**: 441–459.

Cloud, P., and Morrison, K., 1979, On microbial contaminants, micropseudofossils, and the oldest records of life, *Precambrian Research* **9**: 81–91.

Schopf, J. W., and Walter, M. R., 1983, Archean microfossils: new evidence of ancient microbes In: Schopf, J. W. (Ed.) *Earth's Earliest Biosphere, Its Origin and Evolution*, Princeton University Press, Princeton, NJ, pp. 214–239.

# Chapter 9
# ~3,460 Ma Apex Basalt, East Pilbara, Western Australia

**Fig. B22** The Apex Basalt is a 1.5–2.0 km thick formation consisting of tholeiitic pillow lava (*upper* image), high-magnesium basalt, and komatiite, inter-bedded with thinner (<10 m thick) chert members (*lower* image). Putative (and some of the most famous and controversial) microfossils have been reported from the vicinity of the lowermost chert member at Chinaman Creek in the Marble Bar greenstone belt. Scale: hammer is 40 cm long (Photographs by the author)

D. Wacey, *Early Life on Earth: A Practical Guide*,

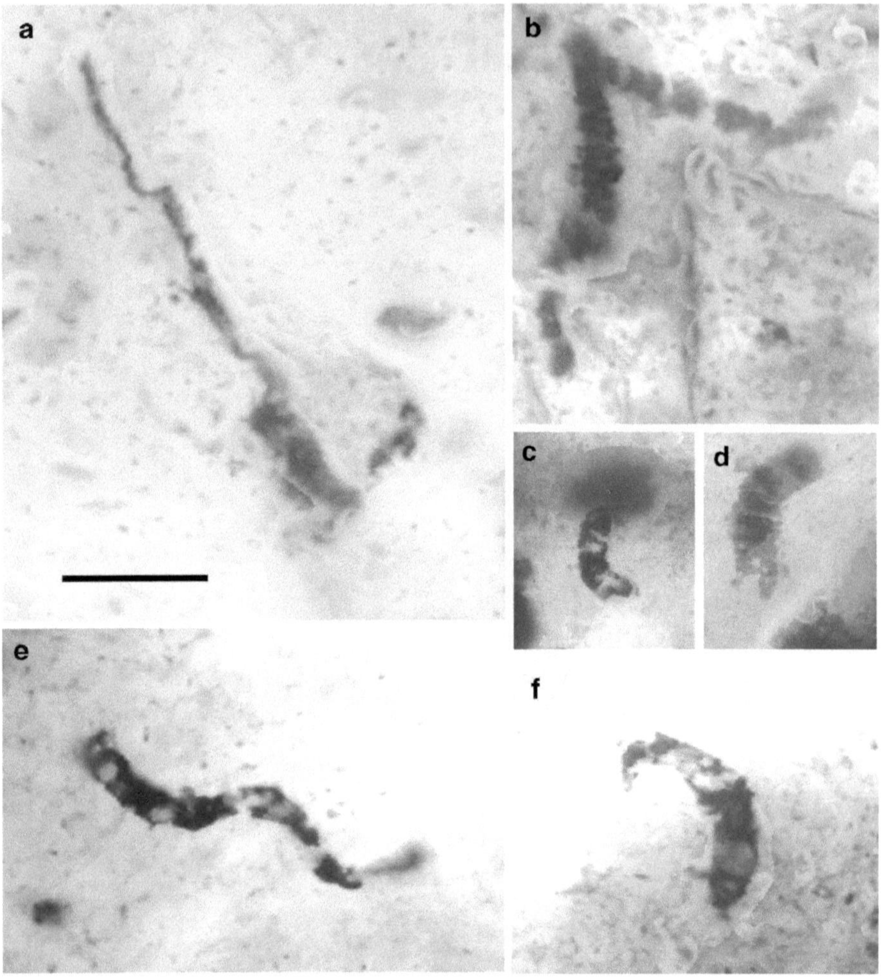


**Fig. B23** Controversial microfossil-like structures from a chert breccia vein within the Apex Basalt (6 out of 11 putative species pictured here). These structures have either been interpreted as microfossils (Schopf, 1992, 1993, 1999, 2006; Schopf et al., 2002, 2007; Schopf and Packer, 1987) or as non-biological artefacts (see pages 12–15, 157–158 and Brasier et al., 2002, 2005, 2004, 2006, 2008).
(**a**) ***Eoleptonema apex***
(**b**) ***Archaeoscillatoriopsis disciformis***
(**c**) ***Primaevifilum laticellulosum***
(**d**) ***Archaeoscillatoriopsis maxima***
(**e**) ***Primaevifilum amoenum***
(**f**) ***Primaevifilum conicoterminatum***
Scale bar is about 15 μm for (**a**) and (**b**); 40 μm for (**c**); 60 μm for (**d**); 20 μm for (**e**) and (**f**) (Photographs courtesy of Martin Brasier)

| | |
|---|---|
| Age of host rock? | ~3,460 Ma: Apex Basalt, Marble Bar greenstone belt, Pilbara, Western Australia; age constrained by U–Pb zircon dates of 3,458 ± 2 Ma from overlying Panorama Fm and 3,471 ± 5 Ma from underlying Duffer Fm |
| Age of structures? | Debatable: Occur in a breccia within a hydrothermal vein system; some occur in rounded or angular clasts but many others occur within multiple generations of fissure- and void-filling silica and in the chert matrix |
| Geological context? | Hydrothermal setting; possibly still viable for (hyper)thermophilic life (but see below) |
| Biological morphology? | **FOR:** Three dimensional structures; filamentous, sinuous, rounded apices and often septate; 11 'species' varying in average diameter from 0.5 to 16.5 µm and 28-89 µm in length; bifurcated 'cells' and 'cell' pairs in process of division reported; similarity to modern taxa<br>**AGAINST:** Many are C, J and L shaped structures around spherulitic silica crystals; branches of different diameter; follow ghost rhombic crystal outlines (e.g., Fig. B23b); no clustering; no orientation to bedding; rather large, up to 36 µm in diameter; the reported septa, bifurcated- and paired-'cells' can easily be explained as alternating plates of non-biogenic carbon and micro-quartz grains |
| Biogeochemistry? | **FOR:** Kerogenous; $\delta^{13}C$ bulk carbon average −27.7‰ PDB; correlation of laser Raman chemical images with optical images indicating kerogenous 'cell walls'<br>**AGAINST:** $\delta^{18}O$ from $SiO_2$ ~ +14‰ which suggests hydrothermal temperatures (~250–350°C) in excess of those that life can survive; laser Raman signal is not an indicator of biology – it is merely an index of preservation; carbon isotope signal has been reproduced in the laboratory when synthesising non-biological carbon |
| Type of organism? | Interpreted as trichomes of bacteria (beggiatoaceans or chloroflexaceans) or cyanobacteria (oscillatoriaceans) by Schopf and Packer (1987), and by Schopf (1993) |
| INTERPRETATION IN THE LITERATURE | 1. Microfossils (see Schopf references)<br>2. Non-biological artefacts (see Brasier references) |
| **IS THIS REALLY LIFE?** | A non-biological origin is most likely (see Fig. B24 and pages 12–15). These are some of the most intensely studied putative Archean microfossils, yet their origin is still very controversial |

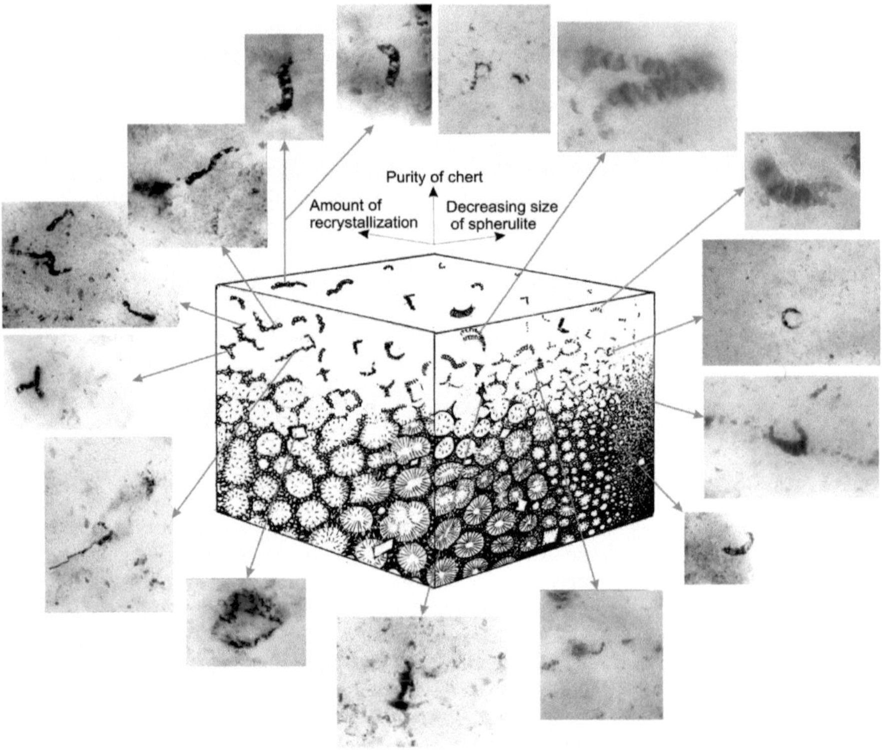


**Fig. B24** A non-biological explanation for the Apex Basalt microstructures. This three dimensional morphospace model (*centre* block) of the Apex 'microfossils' (*outer* images) shows how this spectrum of microfossil-like structures may have been created entirely by physico-chemical controls during recrystallization of the chert and the redistribution of carbonaceous material around spherulite and crystal margins (cf. Brasier et al., 2002, 2004, 2005, 2006, 2008). The key controls here were the relative purity of the chert (*vertical axis*), the degree of recrystallization of the fibrous chalcedony to equigranular microcrystalline chert (*left horizontal axis*), and the decreasing size of the spherulites (*right horizontal axis*). Arrows link theoretical with observed and reported microfossil-like artefacts having similar morphologies (Image courtesy of Martin Brasier)

## Recommended Reading

Brasier, M. D., Green, O. R., Jephcoat, A. P., Kleppe, A. K., Van Kranendonk, M. J., Lindsay, J. F., Steele, A., and Grassineau, N. V., 2002, Questioning the evidence for Earth's oldest fossils, *Nature* **416**: 76–81.

Brasier, M. D., Green, O. R., and Mcloughlin, N., 2004, Characterization and critical testing of potential microfossils from the early Earth: the Apex 'microfossil debate' and its lessons for Mars sample return, *International Journal of Astrobiology* **3**: 1–12.

Brasier, M. D., Green, O. R., Lindsay, J. F., McLoughlin, N., Steele, A., and Stoakes, C., 2005, Critical testing of Earth's oldest putative fossil assemblage from the ~3.5 Ga Apex Chert, Chinaman Creek Western Australia, *Precambrian Research* **140**: 55–102.

Brasier, M. D., McLoughlin, N., and Wacey, D., 2006, A fresh look at the fossil evidence for early Archaean cellular life, *Philosophical Transactions of the Royal Society B* **361**: 887–902.

Brasier, M. D., Green, O. R., Lindsay, J. F., McLoughlin, N., Stoakes, C. A., Brasier, A., and Wacey, D., 2008, Earth's oldest putative fossil assemblage from the ~3.5 Ga Apex chert, Chinaman Creek, Western Australia: a field and petrographic guide, *Special Publication of the Geological Survey of Western Australia.*

Pasteris, J. D., and Wopenka, B., 2002, Images of the Earth's oldest fossils? (discussion and reply), *Nature* **420**: 476–477.

Pasteris, J. D., and Wopenka, B., 2003, Necessary, but not sufficient: Raman identification of disordered carbon as a signature of ancient life, *Astrobiology* **3**: 727–738.

Schopf, J. W., 1992, Paleobiology of the Archaean. In: Schopf, J. W., and Klein, C. (Eds.) *The Proterozoic Biosphere: A Multidisciplinary Study*, Cambridge University Press, Cambridge, pp. 25–39.

Schopf, J. W., 1993, Microfossils of the Early Archaean Apex Chert: new evidence for the antiquity of life, *Science* **260**: 640–646.

Schopf, J. W., 1999, *The Cradle of Life*, Princeton University Press, New York, 367p.

Schopf, J. W., 2006, Fossil evidence of Archean life, *Philosophical Transactions of the Royal Society B* **361**: 869–885.

Schopf, J. W., and Packer, B. M., 1987, Early Archaean (3.3 Billion to 3.5 Billion-year-old) microfossils from Warawoona Group, Australia, *Science* **237**: 70–73.

Schopf, J. W., Kudryavtsev, A. B., Agresti, D. G., Wdowiak, T. J., and Czaja, A. D., 2002, Laser-Raman imagery of Earth's earliest fossils, *Nature* **416**: 73–76.

# Chapter 10
# ~3,450 Ma, Hoogenoeg Formation, Barberton, South Africa

**Fig. B25** The Hoogeneog Formation is up to ~4 km thick and consists largely of massive and pillowed basalts, with a middle komatiitic unit, upper felsic volcanic unit and numerous thin inter-bedded cherts. Putative signs of life have been found in the pillow basalts (*top*) and the thin cherts (*bottom*). Scale: pencil is 15 cm long (Photographs courtesy of Nicola McLoughlin and Martin Brasier)

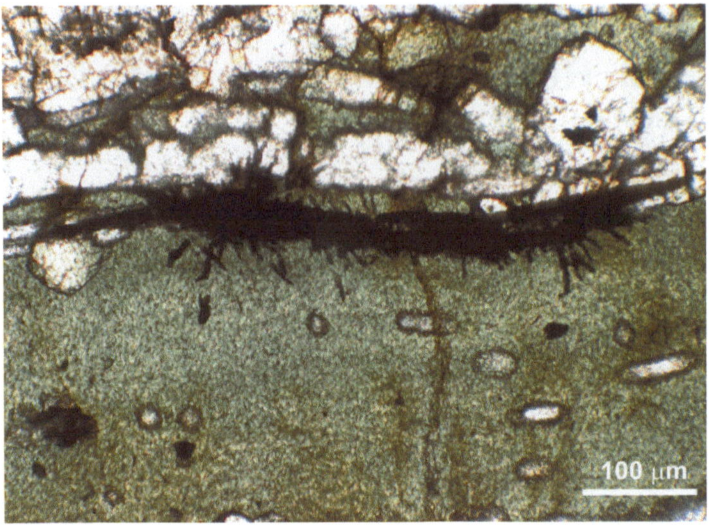


**Fig. B26** Putative trace fossils (dark tubes in centre of image) within a pillow basalt (Photograph courtesy of Nicola McLoughlin)

| | |
|---|---|
| Age of host rock? | 3,480–3,220 Ma: Hooggenoeg and Kromberg Formations, Barberton greenstone belt, South Africa |
| Age of structures? | >3,200 Ma: Interpreted to have formed soon after eruption of the pillow lavas using overlapping metamorphic and magmatic dates |
| Geological context? | Found around rehealed fractures in the formerly glassy margins of pillow basalts erupted on the seafloor |
| Biological morphology? | Micro-tubular. Average 4 μm width and up to 200 μm length; clustered; similar to modern biological analogues |
| Biogeochemistry? | $\delta^{13}C$ bulk carbonate (−16.4‰ to +3.9‰ PDB) from glassy rims compared to crystalline interiors (−6.9‰ to +0.7‰ PDB); possible enrichment of carbon within microtubes |
| Type of organism? | Unknown, thermophilic endolith has been suggested |
| INTERPRETATION IN THE LITERATURE | Trace fossil (Furnes et al., 2004) |
| **IS THIS REALLY LIFE?** | Possible life, although formation mechanism of such microtubes still poorly understood even in modern examples |

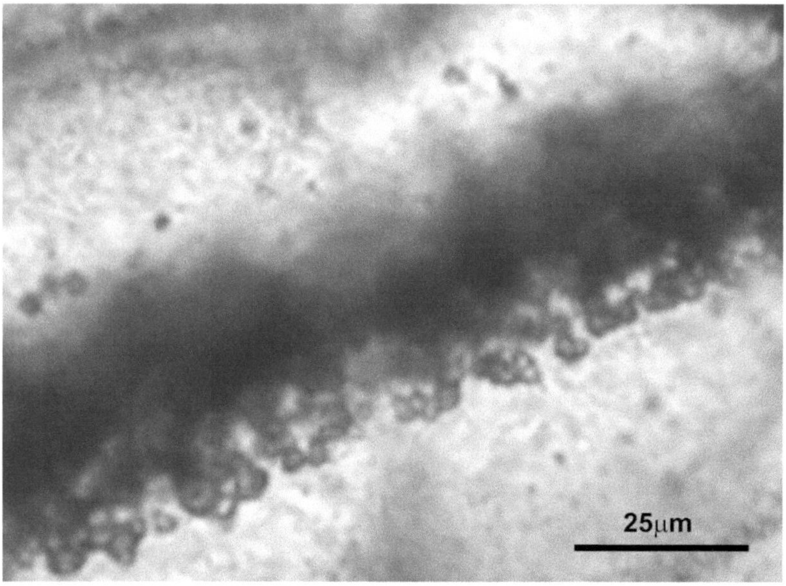


**Fig. B27** Putative granular trace fossils (clusters of spheres just below centre) within a pillow basalt (Photograph courtesy of Nicola McLoughlin)

| | |
|---|---|
| Age of host rock? | 3,472–3,456 Ma, upper Hooggenoeg Formation, Barberton greenstone belt, South Africa |
| Age of structures? | Interpreted to have formed soon after eruption of the pillow lavas using overlapping metamorphic and magmatic dates |
| Geological context? | Found in glass shards within interpillow hyaloclastite, interpreted to have spalled off the glassy margins of pillow basalts |
| Biological morphology? | 1–4 µm diameter spheres, often clustered; close comparison to modern biological weathering |
| Biogeochemistry? | $\delta^{13}C$ bulk carbonate (−16.4‰ to +3.9‰ PDB) from glassy rims compared to crystalline interiors (−6.9‰ to +0.7‰ PDB) |
| Type of organism? | Unknown |
| INTERPRETATION IN THE LITERATURE | Trace fossil (Banerjee et al., 2006) |
| **IS THIS REALLY LIFE?** | Possible life, although formation mechanism of such structures still poorly understood even in modern examples |

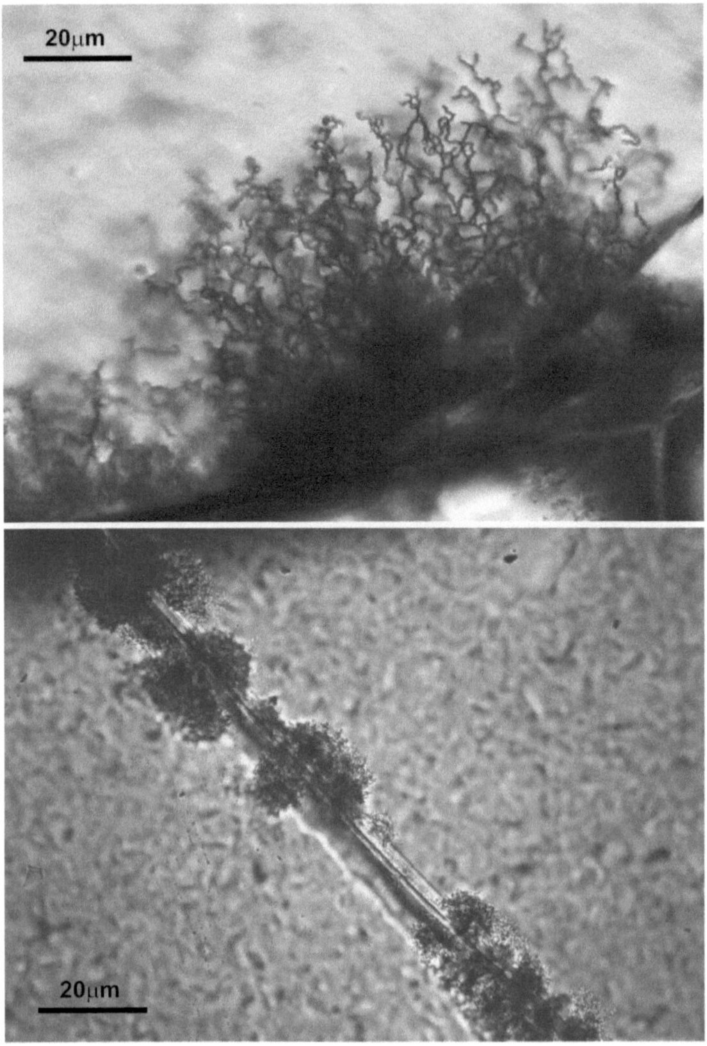


**Fig. B28** Younger analogues for the putative trace fossils described from pillow basalts in the Hoogeneog Formation. *Top*, tubular structures rooted on a fracture in basaltic glass from Quaternary *in situ* oceanic crust (Mid Atlantic Ridge). *Bottom*, granular alteration texture consisting of spheres coalesced along cracks in basaltic glass from 110 Ma *in situ* oceanic crust (ODP 418, Bermuda). Strong evidence for biogenicity comes from nucleic acids that have been identified in both examples (Photographs courtesy of Nicola McLoughlin)

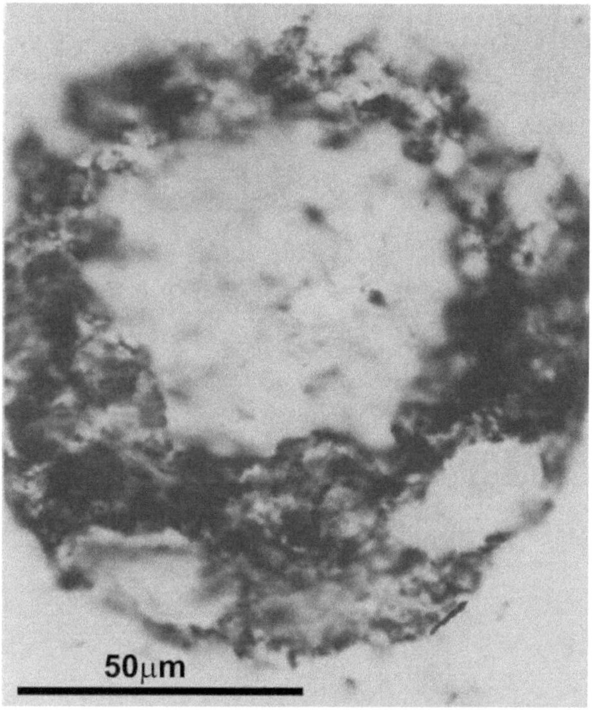


**Fig. B29** Carbonaceous sphere from the Hoogenoeg Formation (Image reproduced from Engel et al., 1968, with permission from AAAS)

| | |
|---|---|
| Age of host rock? | ~3,450 Ma: Hoogenoeg Formation, Barberton greenstone belt, South Africa |
| Age of structures? | ~3,450 Ma: Indigenous to the host cherts, occur in thin sections |
| Geological context? | Poorly described; occur in black carbon-rich cherts inter-bedded with thick lavas |
| Biological morphology? | Poorly described; spheres about 5–105 μm in diameter |
| Biogeochemistry? | Spheres have kerogenous walls |
| Type of organism? | Alga-like (Engel et al., 1968) |
| INTERPRETATION IN THE LITERATURE | 1. Alga-like microfossil (Engel et al., 1968)<br>2. Non-biological sphere (Schopf and Walter, 1983)<br>3. Fossilised gas bubble (Westall et al., 2001) |
| **IS THIS REALLY LIFE?** | Most likely a non-biological artefact in agreement with the Schopf and Walter (1983) interpretation |

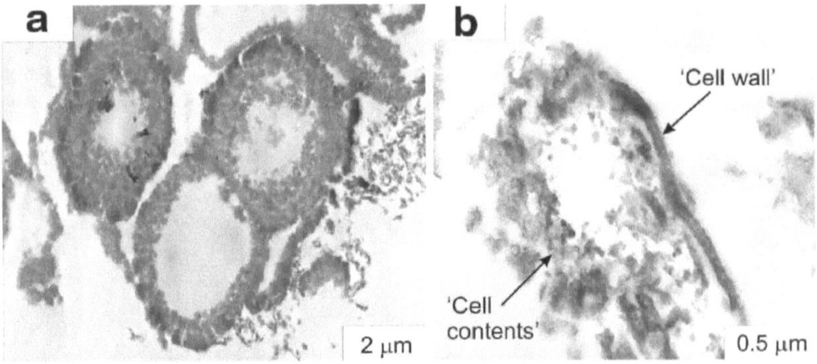


**Fig. B30** Putative remnants of hyperthermophilic microbes within chert from the Hoogenoeg Formation (Image reproduced from Glikson et al., 2008, with permission from Elsevier)

| | |
|---|---|
| Age of host rock? | ~3,450 Ma: Hoogenoeg Formation, Barberton, South Africa; age constrained by U–Pb dates of 3,438 ± 12 Ma and 3,472 ± 5 Ma from zircons at the top and base of the formation respectively<br>Note: Similar structures occur in the ~3,490 Ma Dresser Formation from the Pilbara |
| Age of structures? | Most carbonaceous material occurs as laminations parallel to bedding (likely syndepositional); some occurs dispersed throughout the rock matrix (unknown age) |
| Geological context? | Four types of carbonaceous material occurring in sedimentary bedded cherts; habitat viable for life |
| Biological morphology? | Type A: 2–10 μm aggregates of spherical to sub-spherical bodies with *granular* 'walls' (pictured in (a) above)<br>Type A1: Same as A, but lacking distinct porosity and with 'cell walls' detached (pictured in (b) above)<br>Type B: ~2 μm particles dispersed in rock matrix (no biological morphology)<br>Type C: Clusters of spherical to sub-spherical bodies with *smooth* outer 'walls' preserved in fluid inclusions |
| Biogeochemistry? | None on Barberton samples, but $\delta^{13}C$ of −36.5‰ to −32.1‰ from similar structures in the Dresser Formation samples |
| Type of organism? | Possible hyperthermophile similar to *Methanocaldococcus jannaschii* |
| INTERPRETATION IN THE LITERATURE | Silicified and partially degraded chemosynthetic microbes (Glikson et al., 2008) |
| **IS THIS REALLY LIFE?** | **Presence of degradation features (if interpreted correctly) and close morphological similarity to a degraded modern hyperthermophile are good arguments for biogenicity** |

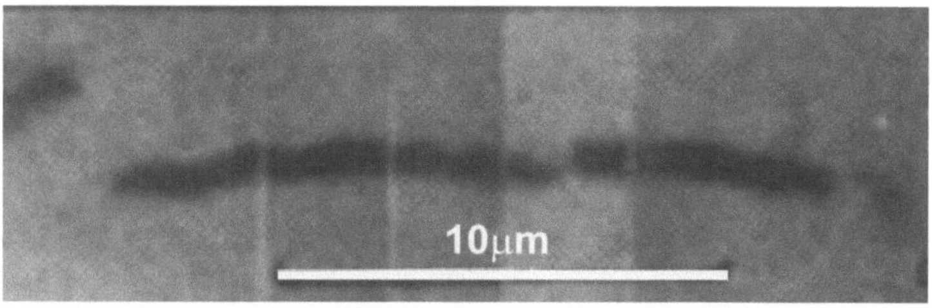


**Fig. B31** Putative microfossil from a banded chert in the Hoogenoeg Formation (Image reproduced from Walsh and Lowe, 1985, with permission from Nature Publishing)

| | |
|---|---|
| Age of host rock | ~3,450 Ma: Hoogenoeg Formation, Barberton, South Africa; age constrained by U–Pb dates of 3,438 ± 12 Ma and 3,472 ± 5 Ma from zircons within felsic volcaniclastic units at the top and base of the formation respectively |
| Age of structures | ~3,450 Ma: Appear integral to the chert host rock; occur associated with fine carbonaceous laminae and detrital carbonaceous grains; occur away from weathering surfaces or veins, and are of similar colour to surrounding organic material |
| Geological context? | Found in carbonaceous, stratiform black and white cherts that have conformable contacts with underlying and overlying sedimentary units; formed during volcanic quiescence in a viable habitat for life; the rocks are of low metamorphic grade |
| Biological Morphology? | Filaments: 0.2–2.6 μm in diameter, up to 200 μm in length; solid, cylindrical and threadlike; non-septate (no evidence for cells) |
| Biogeochemistry? | Filaments composed of carbon and pyrite<br>No other geochemistry reported |
| Type of organism? | Inferred filamentous bacteria or cyanobacteria (Walsh and Lowe, 1985) |
| INTERPRETATION IN THE LITERATURE | Microfossils (Walsh and Lowe, 1985) |
| **IS THIS REALLY LIFE?** | Possible life but more evidence required. Only one poorly illustrated solid example from this stratigraphic unit; lack of geochemistry |

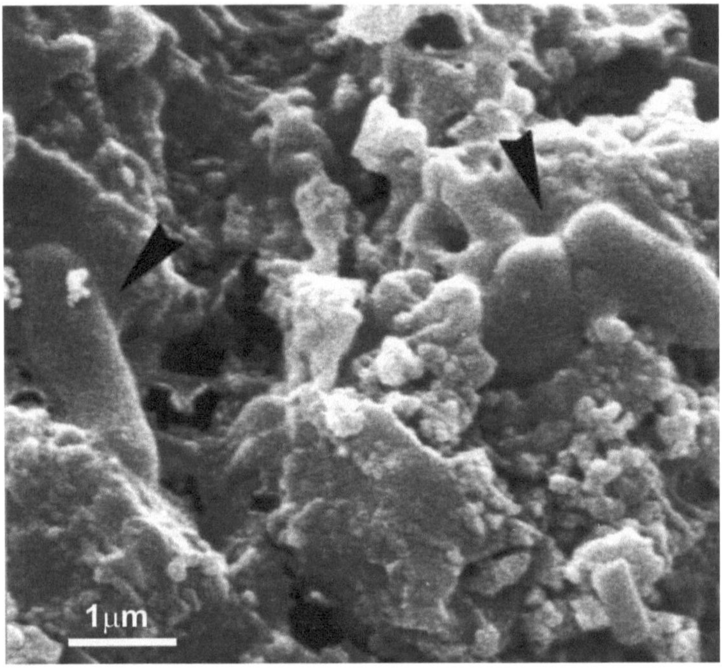


**Fig. B32** Putative sausage shaped microfossils (*arrowed*) from the Hoogeneog Formation (Image reproduced from Westall et al., 2001, with permission from Elsevier)

| | |
|---|---|
| Age of host rock? | ~3,450 Ma: Hoogenoeg Formation, Barberton, SA; age constrained by U–Pb dates of 3,438 ± 12 Ma and 3,472 ± 5 Ma from zircons within felsic volcaniclastic units at the top and base of the formation respectively |
| Age of structures? | ~3,450 Ma: Structures are embedded in a film on a bedding surface of a laminated chert; only seen in etched samples in the SEM, it would be more convincing if they were seen in thin section too |
| Geological context? | Occur within bedded chert interpreted to have formed in shallow marine to sub-aerial environment; viable for life but not all agree on the interpreted environmental setting |
| Biological morphology? | Sausage-shaped (arrowed in image): 2–3.8 μm long and 1.1 μm wide; sometimes joined at their extremities (right hand arrow); can be bent and moulded to the underlying substrate |
| Biogeochemistry? | None |
| Type of organism? | 1. Bacillar bacteria (Westall et al., 2001)<br>2. Sulphate-reducing bacteria (Westall et al., 2006) |
| INTERPRETATION IN THE LITERATURE | Microfossils (Westall et al., 2001, 2006) |
| **IS THIS REALLY LIFE?** | Possible life, but rare and only seen in the SEM – they could possibly be artefacts |

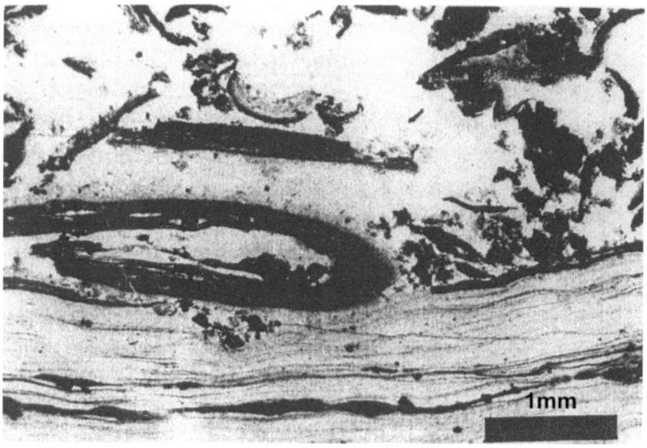


**Fig. B33** Thin section image of a putative microbial mat, part of which is interpreted to have been ripped up and folded over (centre left), from the Hoogenoeg Formation (Image reproduced from Walsh, 1992, with permission from Elsevier)

| | |
|---|---|
| Age of host rock? | ~3,450 Ma: Hoogenoeg Formation, Barberton, South Africa (also found in overlying Kromberg Fm) |
| Age of structures? | ~3,450 Ma: Appear integral to the host chert; within a single traceable layer and away from fractures and veins |
| Geological context? | Found in carbonaceous bedded cherts capping volcanic units – formed during volcanic quiescence; some argument over shallow versus deep water setting |
| Biological Morphology? | 1–20 μm thick laminations that are wavy or crinkly; some are bundled, broken or folded over (pictured above); occur in carbonaceous layers which are 0.5–2 cm thick |
| Biogeochemistry? | Laminations are carbonaceous |
| Type of organism? | Mat building filamentous bacteria or cyanobacteria (Walsh, 1992) |
| INTERPRETATION IN THE LITERATURE | Possible microbial mat (Walsh, 1992) |
| **IS THIS REALLY LIFE?** | Possible life, but non-biological carbon can also form similar structures. Clearer biological evidence needed |

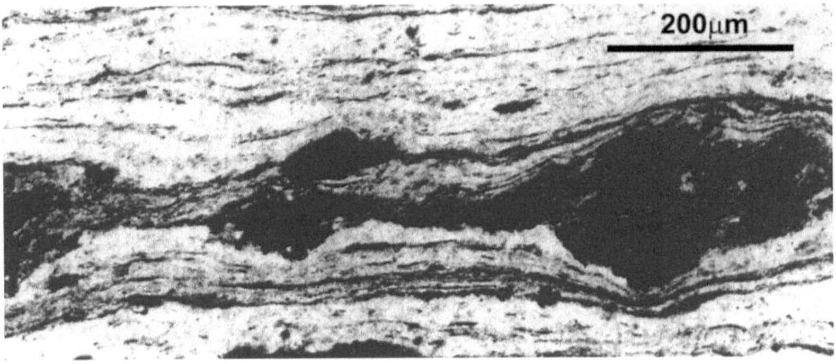


**Fig. B34** Carbonaceous particles from the Hoogenoeg Formation (Image reproduced from Walsh, 1992, with permission from Elsevier)

| | |
|---|---|
| Age of host rock? | ~3,450 Ma: Hoogenoeg Formation, Barberton, South Africa (also found in overlying Kromberg Fm) |
| Age of structures? | ~3,450 Ma: Appear integral to the chert; particles occur aligned parallel to primary carbonaceous laminations |
| Geological context? | Found in carbonaceous bedded cherts capping volcanic units, formed during volcanic quiescence; habitat viable for life |
| Biological morphology? | Individual particles are 25–100 µm in diameter; composite particles are 100–1,000 µm in diameter; associated with laminations (as seen above); not distinctly biological |
| Biogeochemistry? | Carbonaceous |
| Type of organism? | Globular colonies of cyanobacteria (Walsh, 1992) |
| INTERPRETATION IN THE LITERATURE | Microbial material (Walsh, 1992) |
| **IS THIS REALLY LIFE?** | Biological and non-biological explanations are equally compatible with the evidence presented |

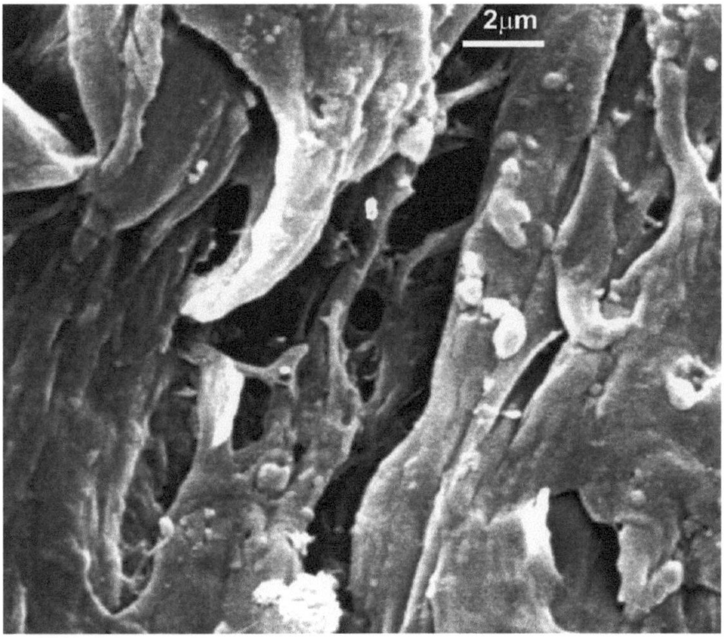


**Fig. B35** SEM image of possible remains of a microbial mat from the Hoogenoeg Formation (Image reproduced from Westall et al., 2001, with permission from Elsevier)

| | |
|---|---|
| Age of host rock? | ~3,450 Ma: Hoogenoeg Formation, Barberton, South Africa; age constrained by U–Pb dates of 3,438 ± 12 Ma and 3,472 ± 5 Ma from zircons within felsic volcaniclastic units at the top and base of the formation respectively |
| Age of structures? | ~3,450 Ma: Granular film-like structure on a bedding plane within a sedimentary bedded chert |
| Geological context? | Bedded chert interpreted to have been deposited in shallow marine to sub-aerial setting; viable for life but not all agree on the interpreted environmental setting |
| Biological morphology? | Micron-sized twisted fibres (pictured above) making up non-rigid structures resembling modern microbially-produced biofilms; lenticular structures are common which are interpreted as gas-filled expansion lenses |
| Biogeochemistry? | None |
| Type of organism? | Unknown mat forming community |
| INTERPRETATION IN THE LITERATURE | Microbial mat (Westall et al., 2001) |
| **IS THIS REALLY LIFE?** | Possible life, but such fibrous material has been formed in non-biological experiments (e.g., Fig. B94) |

**Fig. B36** Filaments and fibrous material representing possible remains of a microbial mat from the Hoogenoeg Formation (Images reproduced from Westall et al., 2006, with permission from The Royal Society)

| | |
|---|---|
| Age of host rock? | ~3,450 Ma: Josefsdal Chert, Hoogenoeg Formation (or possibly lowermost Kromberg Formation), Barberton Greenstone Belt, South Africa. |
| Age of structures? | ~3,450 Ma: Integral part of the host chert; the 'mat' occurs on a bedding plane not a crack; laser Raman micro-spectroscopy shows carbon to be mature kerogen |
| Geological context? | Sedimentary unit that is interpreted to have formed in an evaporitic littoral environment; habitat viable for life although high UV radiation at this time raises questions about survival strategies of organisms in such settings |
| Biological morphology? | 1–4 μm thick 'mat' consisting of multiple layers of parallel and interwoven filaments (e.g., left image) with a near constant diameter of 0.25 μm; filaments are coated by a film that can be ropey, granular, smooth or holey; 'mat' was flexible indicated by flowing movement around detritus (right image) |
| Biogeochemistry? | Bulk carbon isotopes $\delta^{13}C = -22.7‰$ PDB |
| Type of organism? | Anoxygenic photosynthetic filamentous microbes (Westall et al., 2006) |
| INTERPRETATION IN THE LITERATURE | Microbial mat (Westall et al., 2006) |
| **IS THIS REALLY LIFE?** | Possible life, but heavily reliant on morphology; carbon isotope data would be more reliable if measured *in situ* |

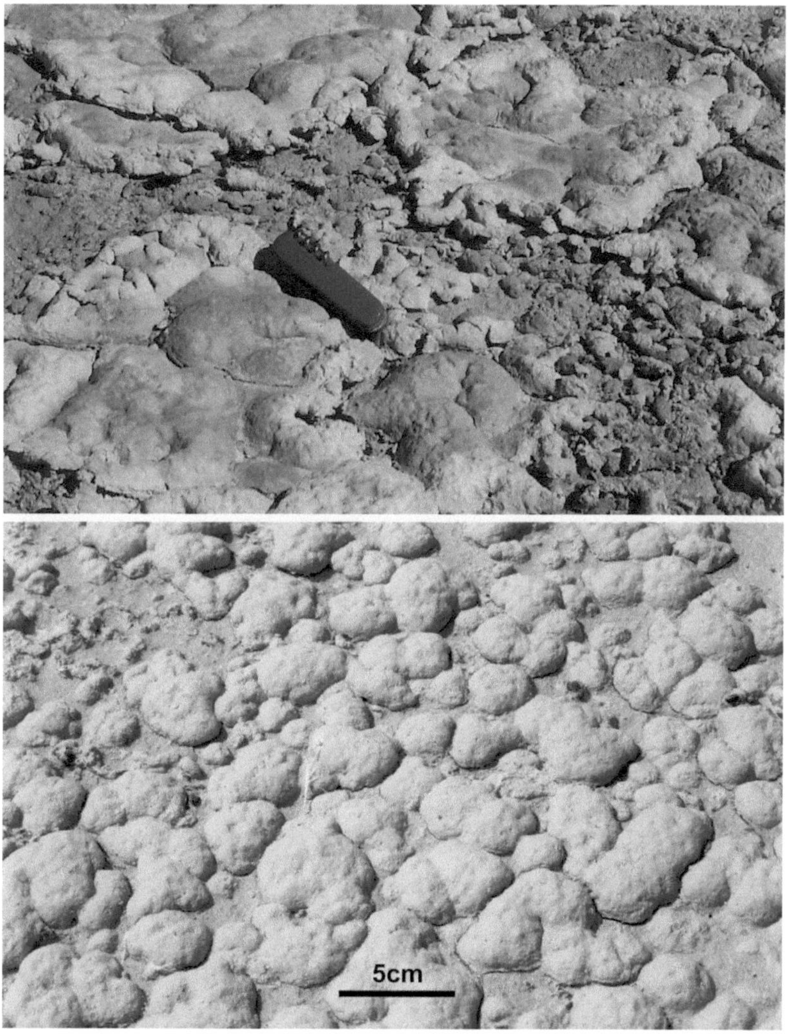


**Fig. B37** Partial modern analogues for the putative microbial mat structures described by Walsh (1992) and Westall et al. (2001, 2006). *Top*, partly desiccated pustular microbial mat from an ephemeral hypersaline lake in the Coorong region of South Australia. *Bottom*, bulbous microbial mat from the shoreline of a second ephemeral lake, Coorong, South Australia (Photographs by the author)

## Recommended Reading


Banerjee, N. R., Furnes, H., Muehlenbachs, K., Staudigel, H., and de Wit, M., 2006, Preservation of 3.4–3.5 Ga microbial biomarkers in pillow lavas and hyaloclastites from the Barberton Greenstone Belt, South Africa, *Earth and Planetary Science Letters* **241**: 707–722.

Engel, A. E. J., Nagy, B., Nagy, L. A., Engel, C. G., Kremp, G. O. W., and Drew, C. M., 1968, Algal-like forms in Onverwacht Series, South Africa: oldest recognised lifelike forms on Earth, *Science* **161**: 1005–1008.

Fisk, M. R., Giovannoni, S. J., and Thorseth, I. H., 1998, The extent of microbial life in the volcanic crust of the ocean basins, *Science* **281**: 978–979.

Furnes, H., and Staudigel, H., 1999, Biological mediation in ocean crust alteration: how deep is the deep biosphere?, *Earth and Planetary Science Letters* **166**: 97–103.

Furnes, H., Banerjee, N. R., Muehlenbachs, K., Staudigel, H., and de Wit, M., 2004, Early Life recorded in Archean pillow lavas, *Science* **304**: 578–581.

Furnes, H., Banerjee, N. R., Muehlenbachs, K., and Kontinen, A., 2005, Preservation of biosignatures in the metaglassy volcanic rocks from the Jormua ophiolite complex, Finland, *Precambrian Research* **136**: 125–137.

Glikson, M., Duck, L. J., Golding, S. D., Hofmann, A., Bolhar, R., Webb, R., Baiano, J. C. F., and Sly, L. I., 2008, Microbial remains in some earliest Earth rocks: comparison with a potential modern analogue, *Precambrian Research* **164:** 187–200.

Nagy, B., and Nagy, L. A., 1969, Early Precambrian Onverwacht microstructures: possibly the oldest fossils on Earth?, *Nature* **223**: 1226–1229.

Schopf, J. W., and Walter, M. R., 1983, Archean microfossils: new evidence of ancient microbes. In: Schopf, J. W. (Ed.) *Earth's Earliest Biosphere, Its Origin and Evolution*, Princeton University Press, Princeton, NJ, pp. 214–239.

Walsh, M. M., 1992, Microfossils and possible microfossils from the early Archean Onverwacht Group, Barberton Mountain Land, South Africa, *Precambrian Research* **54**: 271–293.

Walsh, M. M., and Lowe, D. L., 1985, Filamentous microfossils from the 3,500 Myr-old Onverwacht Group, Barberton Mountain Land, South Africa, *Nature* **314**: 530–532.

Westall, F., de Witt, M. J., Dann, J., van der Gaast, S., de Ronde, C. E. J., and Gerneke, D., 2001, Early Archean fossil bacteria and biofilms in hydrothermally-influenced sediments from the Barberton greenstone belt, South Africa, *Precambrian Research* **106**: 93–116.

Westall, F., de Ronde, C. E. J., Southam, G., Grassineau, N., Colas, M., Cockell, C., and Lammer, H., 2006, Implications of a 3.472–3.333 Gyr-old subaerial microbial mat from the Barberton greenstone belt, South Africa for the UV environmental conditions on the early Earth, *Philosophical Transactions of the Royal Society* B **361**: 1857–1875.

# Chapter 11
# ~3,450 Ma, Panorama Formation, East Pilbara, Western Australia

**Fig. B38** The Panorama Formation consists of felsic volcaniclastic rocks (*upper* image, often heavily sheared and folded – not to be confused with stromatolites!) with minor felsic and mafic lavas, and occasional chert. Evidence for life has been reported from a chert unit (*lower* image) in the Coppin Gap greenstone belt, East Pilbara. Scale: top, pencil is 20 cm long; bottom, pen is 15 cm long (Photographs by the author)

D. Wacey, *Early Life on Earth: A Practical Guide*,

**Fig. B39** Putative microfossils from the Panorama Formation as viewed in the SEM (Image reproduced from Westall et al., 2006, under the fair usage policy of the Geological Society of America)

| | |
|---|---|
| Age of host rock? | ~3,450 Ma: 'Kitty's Gap Chert', Panorama Formation, Coppin Gap greenstone belt, East Pilbara; age constrained from U–Pb date of 3,445 ± 4 Ma from felsic unit within the formation |
| Age of structures? | ~3,450 Ma: Occur within a laterally extensive package of volcaniclastic sediments; spheres are found coating the surface of primary volcanic clasts and are not found in later cavity fill phases |
| Geological context? | Occur in silicified volcaniclastic near-shore sediments. Volcanic clasts are mineral nutrient rich and are a viable substrate for life |
| Biological morphology? | Spheres: Two size ranges, 0.4–0.5 μm (*thin arrow*) and 0.75–0.8 μm (*thick arrow*) in diameter; numerous and often colonial; chains of spheres common; embedded in silicified film interpreted to be microbial extracellular polymeric substances (EPS).<br>Dashed arrow indicates putative desiccation crack<br>Rare rods: Around 1 μm in length |
| Biogeochemistry? | Carbonaceous; bulk $\delta^{13}C$ = –30‰ to –26‰ PDB; Raman spectra consistent with known metamorphic grade |
| Type of organism? | Interpreted to be chemolithotrophs (Westall et al., 2006) |
| INTERPRETATION IN THE LITERATURE | Coccoid microfossils (Westall et al., 2006) |
| **IS THIS REALLY LIFE?** | Possible life but only seen in etched samples under the SEM. Ideally there should be some evidence for these structures in petrographic thin section too. |

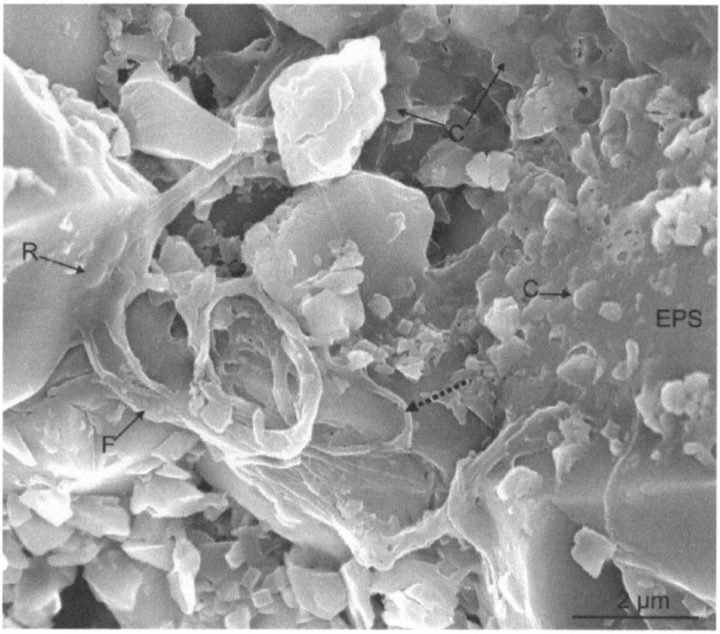


**Fig. B40** Filaments and inferred EPS from the Panorama Formation as seen under the SEM (Image reproduced from Westall et al., 2006, under the fair usage policy of the Geological Society of America)

| | |
|---|---|
| Age of host rock? | ~3,450 Ma: 'Kitty's Gap Chert', Panorama Formation, Coppin Gap greenstone belt, East Pilbara; age constrained by U–Pb date of 3,445 ± 4 Ma from felsic unit within the formation |
| Age of structures? | ~3,450 Ma: Occur within a laterally extensive package of volcaniclastic sediments; found coating the surface of primary volcanic clasts and are not found in later cavity fill phases |
| Geological context? | Occur in silicified volcaniclastic near-shore sediments; volcanic clasts are mineral nutrient rich and are a viable substrate for life; different putative microbial morphologies occur in different layers of the chert suggesting (micro)environmental control |
| Biological morphology? | Filaments (F and dashed arrow): 0.3 μm modal diameter and up to 40 μm in length; associated with spheres (C), rods (R) and inferred EPS; many show degradation features e.g., collapse of cell wall |
| Biogeochemistry? | Carbonaceous; bulk $\delta^{13}C = -30‰$ to $-26‰$ PDB; Raman spectra consistent with known metamorphic grade |
| Type of organism? | Interpreted to be anoxygenic photosynthesisers (Westall et al., 2006) |
| INTERPRETATION IN THE LITERATURE | Microfossils and EPS (Westall et al., 2006) |
| **IS THIS REALLY LIFE?** | A good candidate for life with many supporting lines of evidence – unfortunately the structures are only seen in etched samples under the SEM. Would like to see them in petrographic thin section too |

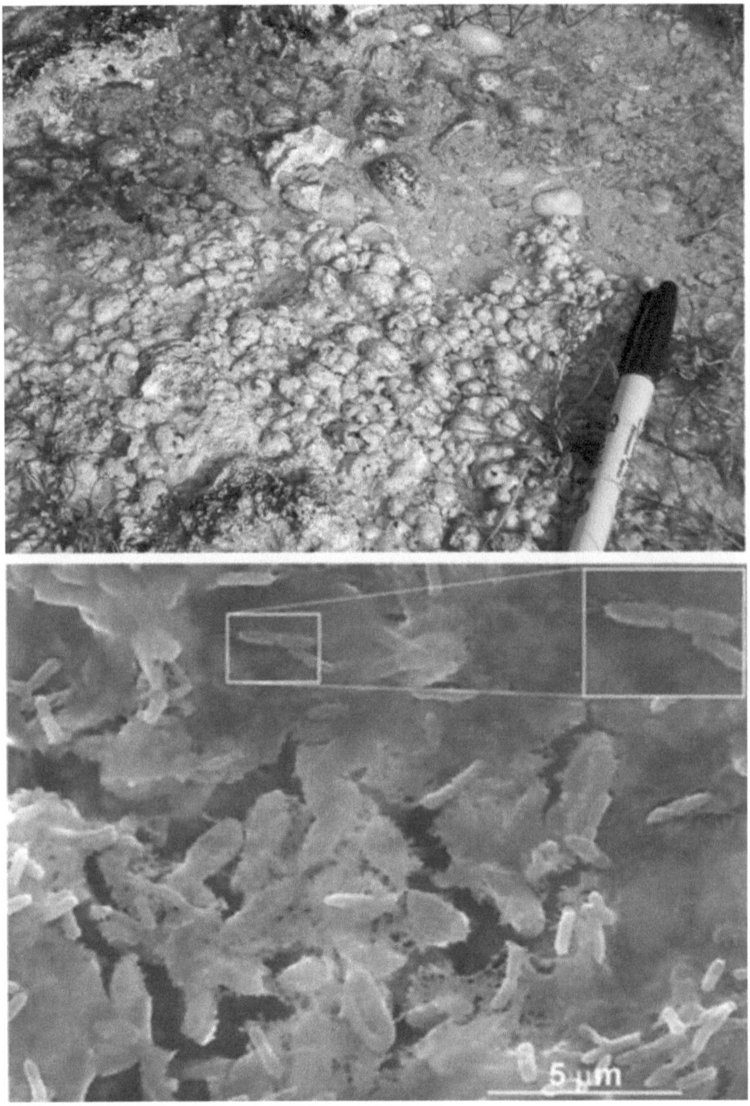


**Fig. B41** Modern analogues for the putative microbial mat and microfossils from the Panorama Formation. *Top*, modern pustular microbial mat from Lake Clifton, Western Australia (Photograph by the author). *Bottom*, SEM image of part of a modern marine biofilm consisting of extracellular polymeric substances and rod-like cells (Toporski et al., 2002, reproduced with permission from Elsevier)

## Recommended Reading


Orberger, B., Rouchon, V., Westall, F., de Vries, S. T., Pinti, D. L., Wagner, C., Wirth, R., and Hashizume, K., 2006, Microfacies and origin of some Archean cherts (Pilbara, Australia), *GSA Special Paper* **405**: 133–156.

Rouchon, V., Pinti, D. L., Gallien, J. P., Orberger, B., Daudin, L., and Westall, F., 2005, NRA analyses of N and C in hydromuscovite aggregates from a 3.5 Ga chert from Kitty's Gap, Pilbara, Australia, *Nuclear Methods and Instruments in Physics Research* **B231**: 536–540.

Toporski, J. K. W., Steele, A., Westall, F., Avcr, R., Martill, D. M., and McKay, D. S., 2002, Morphologic and spectral investigation of exceptionally well-preserved bacterial biofilms from the Oligocene Enspel Formation, Germany, *Geochimica et Cosmochimica Acta* **66**: 1773–1791.

Walsh, M. M., 2004, Evaluation of early Archean volcanoclastic and volcanic flow rocks as possible sites for carbonaceous fossil microbes, *Astrobiology* **4**: 429–437.

Westall, F., 1999, The nature of fossil bacteria, *Journal of Geophysical Research of Planets* **104**: 16437–16451.

Westall, F., and Walsh, M. M., 2003, Fossil biofilms and the search for life on Mars. In: Krumbein, W. E. et al. (Eds.) *Fossil and Recent Biofilms*, Kluwer, Amsterdam, pp 447–465.

Westall, F., de Vries, S. T., Nijman, W., Rouchon, V., Orberger, B., Pearson, V., Watson, J., Verchovsky, A., Wright, I., Rouzaud, J. N., Marchesini, D., and Severine, A., 2006, The 3.446 Ga "Kitty's Gap Chert", an early Archean microbial ecosystem, *GSA Special Paper* **405**: 105–131.

# Chapter 12
# ~3,426–3,350 Ma, Strelley Pool Formation, East Pilbara, Western Australia

**Fig. B42** The Strelley Pool Formation is a widespread sedimentary deposit occurring as a marker horizon across numerous East Pilbara greenstone belts. It generally consists of a basal silicified sandstone and conglomerate unit (*lower* image), overlain by banded cherts (*upper* image) representing silicified carbonates and volcanic ash. The Formation is often capped by an intra-formational breccia. Putative signs of life have been reported from both the sandstone and chert units. Scale: *top*, hammer tip is 3 cm across; *bottom*, pencil is 15 cm long (Photographs by the author)

D. Wacey, *Early Life on Earth: A Practical Guide*,

## 12.1 Summary of Claims of Early Life from this Formation

The Strelley Pool Formation is the rock unit that I myself have studied in most detail. I will now take this opportunity to summarise the various claims for life from this Formation; some are widely accessible in the literature, whilst others are new discoveries. Beginning at the base of the Formation, the Strelley Pool sandstone contains well rounded detrital grains of pyrite, chromite, rutile, and zircon, occurring as a heavy mineral placer-type deposit within a beach (or plausibly even river) setting. Many of the pyrite grains contain microborings (Fig. B43). These microborings typically comprise meandering cylindrical tunnels (microtubes) or channels, or spherical to elliptical surface pits. They have a near constant diameter of 1–5 µm and can penetrate into the substrate for several tens of microns. They are often concentrated in clumps along one side of a mineral grain. Some microborings show slight constrictions along their margins suggestive of colonisation by individual cells (Fig. B43). The microborings do not extend into adjacent grains nor into the microcrystalline silica cement. Nano-scale geochemical mapping shows enrichments of carbon and nitrogen within and/or along the edges of the microborings, consistent with some preservation of biological material (Wacey et al., 2008c).

Non-biological features which closely resemble microborings include microfractures and physico-chemical erosion and corrosion marks (Fig. B102). Microfracturing could be rejected as a formation mechanism for these microborings because they lack a jigsaw puzzle fit, are not sheet-like, and show a nearly constant diameter. Their origin from erosion and corrosion structures was also rejected because the microtubes penetrate far into the substrate and most are an order of magnitude smaller than co-occurring pit marks. Ambient inclusion trails can also result in microtubular structures but these would be expected to have striations along their length, polygonal cross sections and show evidence for a propelled crystal at their terminations (see Fig. B45).

The colonisation of sandstones by endolithic bacteria is a common process at the present day. Endolithic bacteria can tolerate a wide range of ecological conditions and have been shown to inhabit some of the most hostile environments on Earth. Indeed, microborings created by modern endolithic iron and sulphur-oxidising microbes show close similarity to those found in the Strelley Pool sandstone (Fig. B44). Such endolithic microorganisms therefore appear to be logical candidates for preservation of the earliest life forms because this mode of life is well adapted to protection against the harsh surface conditions, such as high UV radiation and meteorite impacts upon the early Earth.

Much interest has surrounded the discovery of endolithic bacteria in Antarctic sandstones from the Ross Desert (e.g., Friedmann and Weed, 1987; Ascaso and Wierzchos, 2002). Here, the mean annual temperature is minus 25°C, meaning the metabolic activity of these endoliths can only be sustained for very short periods of time. Despite this, cyanobacteria, fungi and proto-lichens exist within the upper portions of translucent sandstones and in cracks in granitic rocks. It has been suggested that these extremophile organisms could provide clues to the type of life (if ever present) on Mars.

A second type of microtubular structure from the Strelley Pool sandstone occurs in well rounded, detrital, microcrystalline *silica* grains. These structures are strictly substratum specific, confined to those clasts of well-preserved microcrystalline silica that contain small cubes of pyrite. The microtubes are 1–15 μm in diameter, up to 300 μm in length and can either be hollow or infilled with silica, ferrous phosphate, aluminium phosphate or jarosite (Fig. B45). They exhibit a range of morphologies from straight, to curved, twisted or even helical, and are occasionally branched. Their diameter appears to remain constant along their length, and they commonly cluster around clots of organic matter. These have been interpreted as ambient inclusion trails (AIT; see page 29) for the following reasons (Wacey et al., 2008a):

- They only occur in clasts rich in pyrite (or pseudomorphs after pyrite) and commonly also rich in organic matter.
- Some still contain terminal pyrite crystals.
- Many microtubes originate from the centre of the silica grains and can move both inwards and outwards through the grain. Endolithic borings would only penetrate inwards from the edge of a grain.
- Many of the microtubes have polygonal cross sections, consistent with the geometry of a mineral grain, but inconsistent with microfossils or microbial borings.

Nano-scale biological signals found within these AIT from the Strelley Pool sandstone include carbon and nitrogen enrichment, carbon isotope signal consistent with biology, enrichments of biologically important trace elements (e.g., Ni, Co, Zn). However, as with all such discoveries, Wacey et al. (2008a) caution that contamination from later fluids must be discounted before AIT in general can be used as a biomarker in ancient rocks. Ongoing studies using younger examples of AIT (Fig. B46; Wacey et al., 2008b) hope to shed more light on the specific formation mechanism of these intriguing structures.

Nano-scale biological signals are also being found associated with tiny pyrite grains within the Strelley Pool sandstone. Here dense clusters of sub-micron- to micron-sized anhedral to subhedral pyrite grains coat larger framework quartz grains (Fig. B48). Thousands of such grains can occur in a single thin section. Many grains are intimately associated with carbon- and nitrogen-rich, dark brown to black organic material interpreted to be biological in origin (Fig. B48). Rounded sandstone intraclasts containing such pyrite occur higher within the sandstone member, supporting petrographic observations that this pyrite formed *in situ* during deposition and early cementation of the sandstone. Importantly, no pyrite is found in sandstone surrounding these intraclasts. Isotopic studies are ongoing, but initial results are suggestive of biological sulphur processing (Wacey and Kilburn, 2008).

Concentrically laminated structures that resemble ooids are also found within the Strelley Pool sandstone (Fig. B51). These are found wholly contained within lithified 'intraclasts' that have subsequently been incorporated into the sandstone/conglomerate. Their age is thus well constrained. The grains are between 300 μm and 3 mm in diameter and range from circular to elliptical or colloform in shape. Several of the grains have nuclei or multiple nuclei and most show well developed concentric or colloform lamination, of potential biogenic origin. They show close resemblance to both Proterozoic and some modern day ooids but are now wholly silificified (see Fig. B52). If these prove to be silicified carbonate ooids, then they will be about a billion years older than other examples. The null hypothesis to be tested here includes their origin as reworked volcanic features such as accretionary lapilli or as reworked amygdales/cavity fills. Work is continuing on these structures.

Moving up in the Formation to the overlying chert unit, a range of stromatolite morphologies have been described from more than one greenstone belt (Lowe, 1980; Hofmann et al., 1999; Allwood et al., 2006, 2007). The best examples come from the Strelley West locality in the East Strelley greenstone belt (Fig. B49) and the intriguing Trendall locality within the North Pole greenstone belt (Fig. B49). Trendall, in particular, exhibits stromatolites notable for possessing a diverse range of coniform and rare columnar morphologies, a significant variation in size, and one example of putative branching (Hoffmann et al., 1999).

The biogenicity of these stromatolites is highly controversial. Initially (Lowe, 1980) biogenicity was inferred based largely on their morphology and comparisons with modern day stromatolites, for example those found at Shark Bay, Western Australia (Fig. B50). Lowe (1994), however, directly questioned his earlier biological interpretation of the Strelley Pool stromatolites, instead concluding that they formed through evaporitic precipitation. The discovery of the Trendall locality structures (Hofmann et al., 1999) led to a biological interpretation being re-advanced for the Strelley Pool stromatolites. This was taken further in a recent study (Allwood et al., 2006) that divided the Strelley Pool stromatolites into seven separate facies and presented a depositional model that is taken to support a shallow water phototropic microbial origin for the stromatolites.

More broadly however, as for example at Strelley West the more typical, small, unbranched coniform stromatolites (Fig. B49a, c, d) do not show the biological characteristics of depth controlled distribution or changes in morphology with depth. In addition, their close interrelationship with crystal fan arrays upon which they can be seen to nucleate emphasises a strong chemical component to their growth. Their intergradation with linguoid current ripples (Fig. B97) highlights the predominance of physical rather than biological processes in their accretion. It is also notable that remains of the microbial mat microfossils inferred to have built these structures, or extensive carbonaceous microfabrics that they may have created, have never been found in association with the stromatolites. However, this absence of evidence may simply be due to the low preservation potential of microfossils in stromatolites given that, even in Phanerozoic stromatolites, microbes are only found in an estimated ~1% of occurrences.

At present, in the absence of supporting microtextural and geochemical evidence, the biogenicity of at least the morphologically less complex stromatolites in this Formation remains to be demonstrated (see also pages 9–12 in the introduction to this book by Brasier).

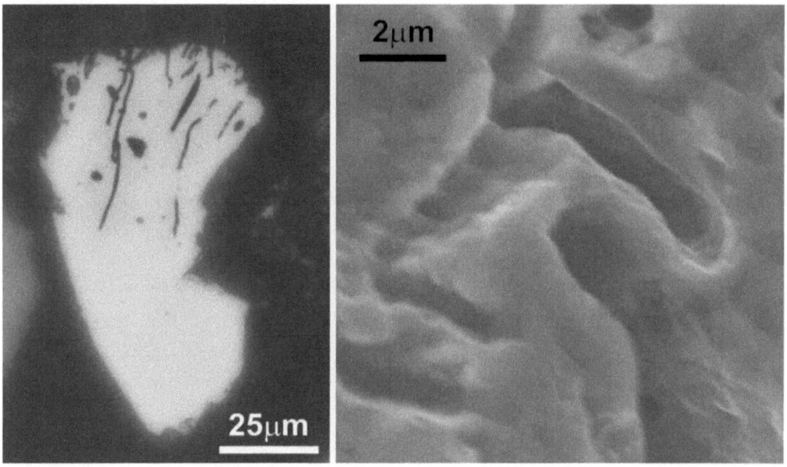


**Fig. B43** Putative trace fossils in pyrite from the Strelley Pool sandstone (Images from the Oxford Palaeobiology Collections)

| | |
|---|---|
| Age of host rock? | ~3,400 Ma: Strelley Pool sandstone, East Strelley greenstone belt, Pilbara, Western Australia; age constrained by U–Pb dates of 3,350 Ma and 3,426 Ma on zircons from overlying and underlying units respectively |
| Age of structures? | ~3,400 Ma: Microborings are found in detrital grains truncated by early silica cement; observed in petrographic thin section, reflected light (above left) and in the SEM (above right) |
| Geological context? | Found in metallic detrital grains (pyrite) within a beach-type sandstone. Viable for life |
| Biological morphology? | 1–5 μm diameter microtubes, channels and indentations (left image); microtube cross sections are spherical to oval; penetrate from outside of grain; often clustered; segmentation of microtubes is common (right image) |
| Biogeochemistry? | High resolution geochemistry shows carbon and nitrogen enrichments within the microtubes |
| Type of organism? | Iron or sulphur oxidising bacteria |
| INTERPRETATION IN THE LITERATURE | Probable trace fossil (Wacey et al., 2008c) |
| **IS THIS REALLY LIFE?** | **A very promising discovery. Work is still in progress to understand the exact metabolisms involved** |

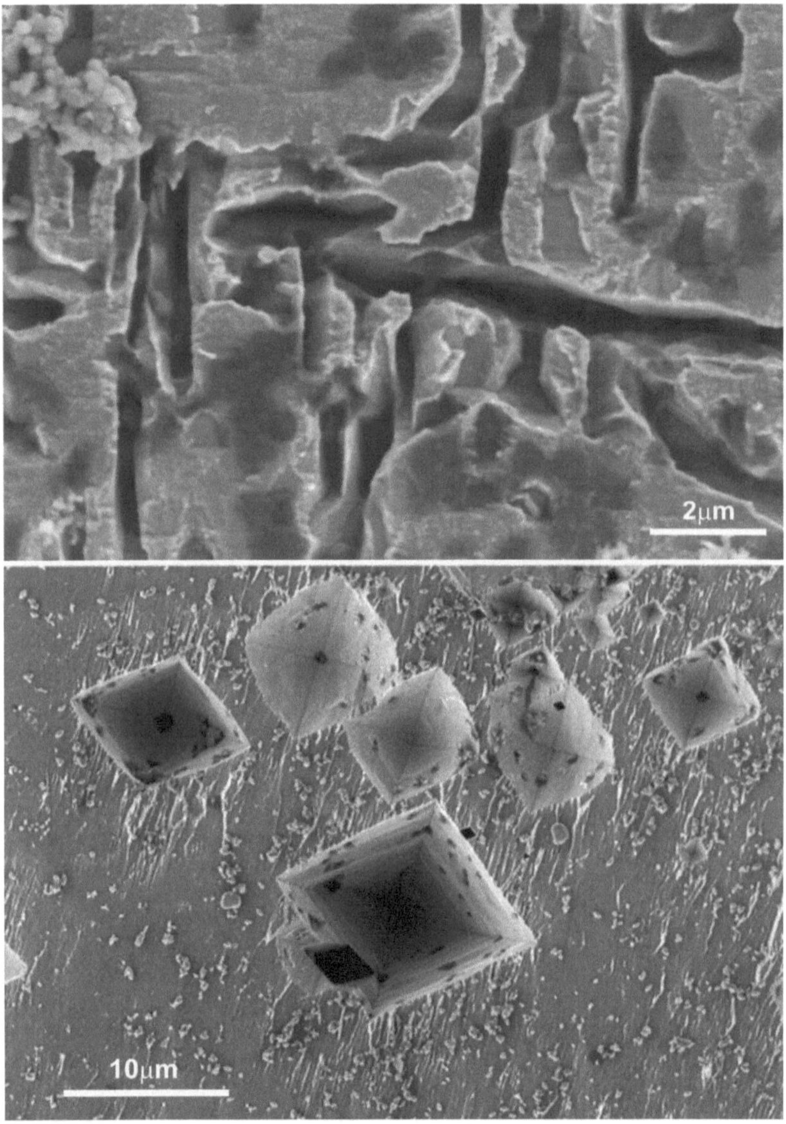


**Fig. B44** Modern laboratory analogues for the putative trace fossils in pyrite grains from the Strelley Pool sandstone. *Top*, channels and pits on a pyrite crystal face, created by the microbial activity of iron-oxidising bacteria. *Bottom*, rhombohedral pits on the surface of a pyrite crystal face, created partly by the microbial activity of iron oxidising archaea but also strongly related to weaknesses in the pyrite crystal lattice structure (hence non-microbial shape) (SEM images courtesy of Katja Etzel)

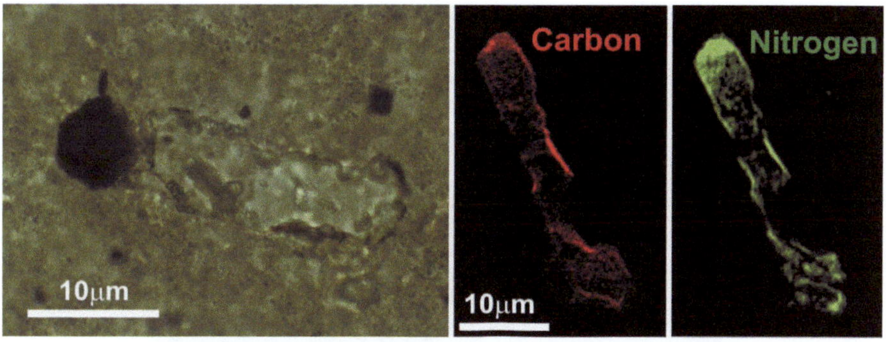


**Fig. B45** Ambient inclusion trails within the Strelley Pool sandstone (Images modified from Wacey et al., 2008a)

| | |
|---|---|
| Age of host rock | ~3,400 Ma: Strelley Pool sandstone, East Strelley greenstone belt, Pilbara, Western Australia; age constrained by U–Pb dates of 3,350 Ma and 3,426 Ma on zircons from overlying and underlying units respectively |
| Age of structures | Partly >3,400 Ma: The pyrite and associated organic material was integral to the detrital grains<br>Partly >2,900 Ma: The tubular trails themselves may have formed anytime during diagenesis and low grade metamorphism |
| Geological context? | Organic material found attached to pyrite grains and lining tubular trails within a beach sandstone; habitat viable for life |
| Biological morphology? | Tubular trails: 1–15 μm in diameter, up to 300 μm in length; can be straight (left image), curved, twisted or helical; have constant diameter and are sometimes branched; often cluster around clots of organic material. Nothing *uniquely* biological |
| Biogeochemistry? | Carbon and nitrogen enrichments within tubular trails (right image) and around pyrite grains; enrichments of other biologically important elements such as Co, Fe, K, Ni, S, Zn within tubular trails; *in situ* $\delta^{13}C = -26‰$ PDB of carbon lining the trails |
| Type of organism? | Unknown, but probable sulphur based metabolism |
| INTERPRETATION IN THE LITERATURE | Ambient inclusion trails catalysed by biology (Wacey et al., 2008a) |
| **IS THIS REALLY LIFE?** | **A comprehensive set of geochemical evidence. More work needed to see if all ambient inclusion trails form in this way** |

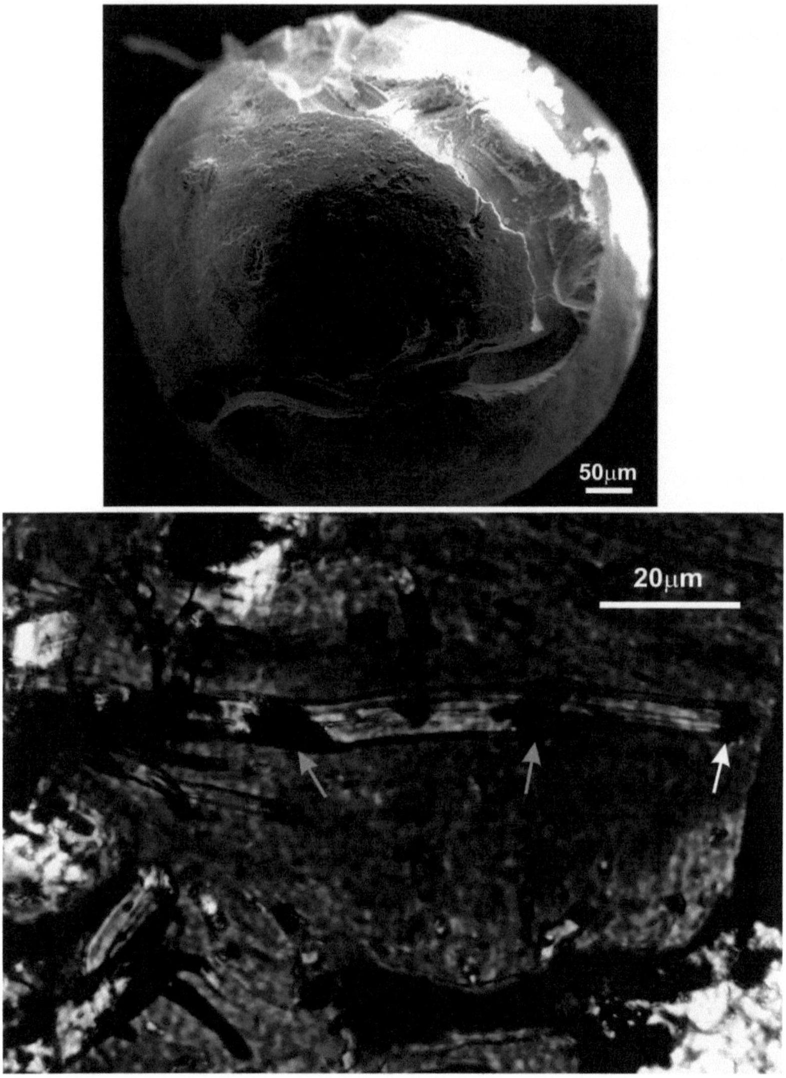


**Fig. B46** Younger analogues for the ambient inclusion trails (AIT) described from the Strelley Pool sandstone. *Top*, large AIT in fossilised 'embryo' from the ~570 Ma Doushantuo Formation, China. *Bottom*, AIT in a fish scale from ~390 Ma Achanarras Limestone, Scotland. A cluster of trails is observed on the left side of the image, with one long example showing longitudinal striations, clumps of organic material (*blue arrows*) and part of a terminal pyrite crystal (*white arrow*) (Images modified from Wacey et al., 2008b)

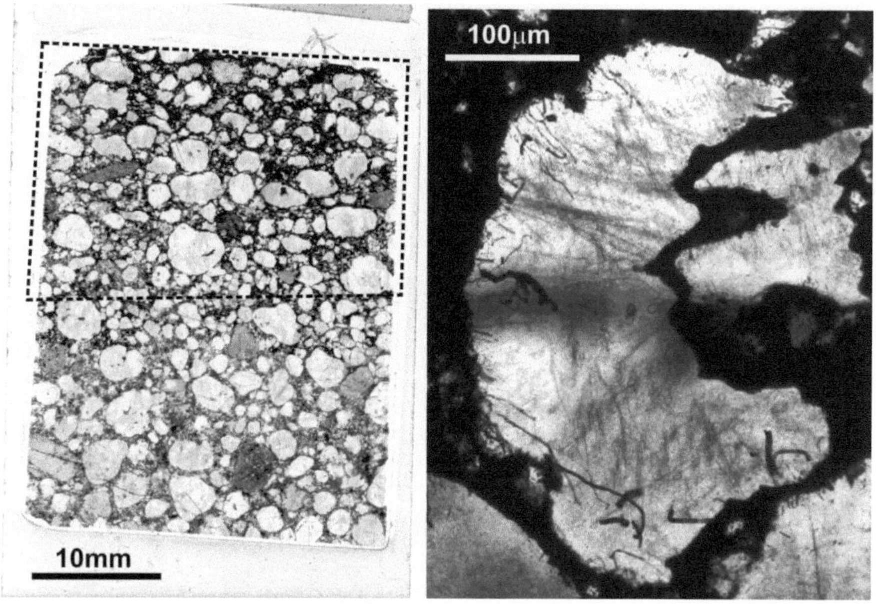


**Fig. B47** Modern endolithic contamination in the Strelley Pool sandstone (Images modified from Wacey et al., 2008b)

| | |
|---|---|
| Age of host rock? | ~3,400 Ma: Strelley Pool sandstone, East Strelley greenstone belt, Pilbara, Western Australia |
| Age of structures? | Modern: Tubular structures cut across all grain types and occur only in outer weathered margin of thin sections (left image); they are filled with iron oxide and hydrous phases that are unlikely to have survived Archean metamorphism |
| Geological context? | Invade weathered surfaces of sandstone in a desert-like environment; harsh but viable for life |
| Biological morphology? | Filaments and microtubes: 1–20 μm in diameter, up to ~500 μm long; tangled and clustered masses within the matrix of the sandstone; also penetrate quartz grains from edge inwards |
| Biogeochemistry? | Carbonaceous; phosphate, sulphur and nitrogen rich |
| Type of organism? | Probably fungal hyphae (Wacey et al., 2008b) |
| INTERPRETATION IN THE LITERATURE | Modern contaminant microfossil and/or trace fossil (Wacey et al., 2008b) |
| **IS THIS REALLY LIFE?** | **Yes, but it is not old!** **Modern contamination is a major problem when investigating early life. Great care and attention to detail is required to eliminate it** |

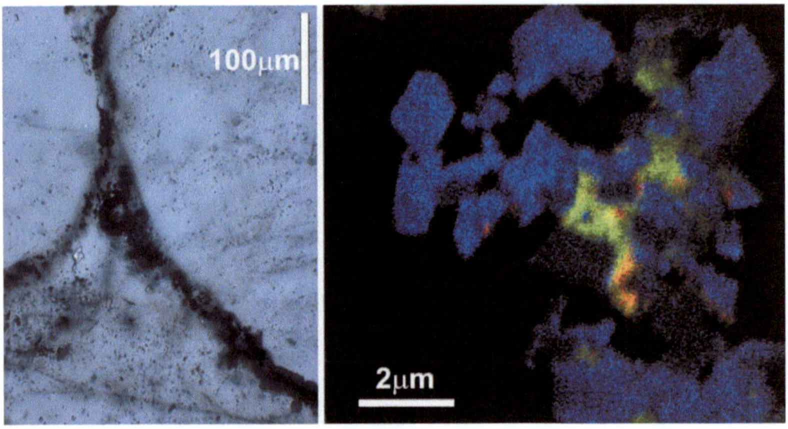


**Fig. B48** Possible biominerals from the Strelley Pool sandstone (Images modified from Wacey and Kilburn, 2008) (see table below for key to colours)

| | |
|---|---|
| Age of host rock | ~3,400 Ma: Strelley Pool sandstone, East Strelley greenstone belt, Pilbara, Western Australia; age constrained by U–Pb dates of 3,350 and 3,426 Ma on zircons from overlying and underlying units respectively |
| Age of structures | >3,350 Ma: Syn-depositional to early diagenetic; rounded intra-clasts of the pyrite-rich sandstone are found higher in the sandstone unit |
| Geological context? | Beach-like setting; pyrite coats framework quartz grains (left image); habitat viable for life |
| Biological morphology? | Pyrite is sub-micron- to micron-sized; organic material is intimately associated with the pyrite; grains cluster and coat framework quartz much as a biofilm might be expected to do |
| Biogeochemistry? | *In situ* $\delta^{34}S$ ranges from −10‰ to +10‰ CDT – a greater range than most non-biological processes; carbon (red) and nitrogen (green) enriched material attached to some pyrites (blue) as shown in the right hand image |
| Type of organism? | Sulphate-reducing microbes; possible sulphur oxidising and disproportionating microbes |
| INTERPRETATION IN THE LITERATURE | Possible biominerals and associated biological material (Wacey and Kilburn, 2008) |
| **IS THIS REALLY LIFE?** | **Possible life, but isotopic range can also be produced without the aid of biology** |

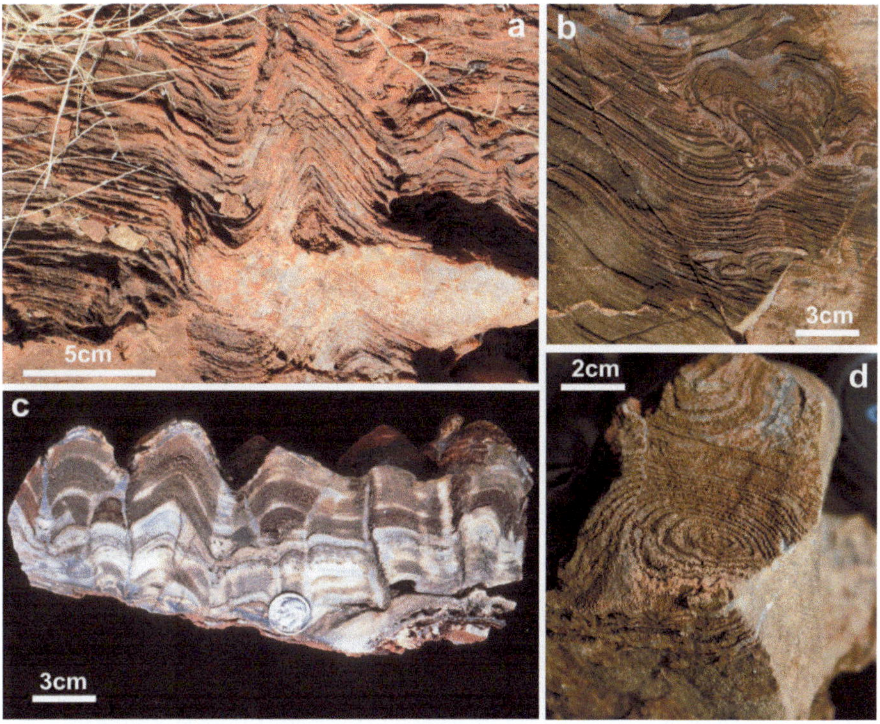


**Fig. B49** Stromatolites from the Strelley Pool chert: (**a**) Cross section through a highly silicified, steep-sided coniform stromatolite (**b**) Cross section through a small double branched stromatolite known as 'Mickey Mouse Ears' (**c**) Cut slab through small coniform stromatolites (**d**) Horizontal cross section through a coniform stromatolite with laminations expressed as rings (**a**), (**c**) and (**d**) are from the 'Strelley West locality'; (**b**) is from the 'Trendall locality' (Photographs except (**c**) by the author)

| | |
|---|---|
| Age of host rock? | ~3,400 Ma: Strelley Pool chert, Pilbara, Western Australia; age constrained by U–Pb dates of 3,350 Ma and 3,426 Ma on zircons from overlying and underlying units respectively |
| Age of structures? | ~3,400 Ma: Primary (bio?)sedimentary structures |
| Geological context? | Interpreted environment of deposition is an isolated peritidal carbonate platform (see Allwood et al., 2007 for detailed environmental interpretation); environment viable for life |
| Biological morphology? | Seven types of laminated structure which show some comparison to modern biological stromatolites; morphologies include 'encrusting and domal laminites', 'small crested/conical laminites', 'egg-carton laminites', 'large complex cones', 'cuspate swales', 'wavy laminites', 'iron-rich laminites' (Allwood et al., 2006). Nothing in the morphology that is definitively biological, although conical forms have yet to be produced non-biologically |
| Biogeochemistry? | Rare earth element enrichment in carbonate laminae relative to chert laminae – this is consistent with younger microbial carbonates |
| Type of organism? | No microfossils have been found associated with these structures. From analysis of the palaeo-environment it has been suggested that the organisms would be halophilic and probably phototrophic |
| INTERPRETATION IN THE LITERATURE | 1. Biological stromatolites (Allwood et al., 2006, 2007; Hofmann et al., 1999)<br>2. Non-biological stromatolites (Lowe, 1994) |
| **IS THIS REALLY LIFE?** | **Possible life, but many types of stromatolite-like structures can be produced without biology (e.g., Fig. B96). Conical stromatolites grade into purely sedimentary structures such as linguoid ripples in some greenstone belts (Fig. B97)** |

**Fig. B50** Younger analogues for the stromatolites described from the Strelley Pool Formation. *Top*, modern columnar stromatolites from a hypersaline lake in the Coorong Region of South Australia. *Bottom*, complex stromatolite showing conical features (*lower right*) from the 2,715 Ma Meentheena carbonate of the Tumbiana Formation, Western Australia (Photographs by the author)

**Fig. B50 (continued)** Younger analogues for the stromatolites described from the Strelley Pool Formation. *Top*, Modern columnar and domal stromatolites from Shark Bay, Western Australia. *Bottom*, spherical microfossils preserved in a Triassic stromatolite from Italy (Perri and Tucker, 2007, image reproduced under the fair usage policy of the Geological Society of America). This is one of the few reported occurrences of microfossils being preserved in stromatolites. Unfortunately microfossils have yet to be found associated with any early Archean stromatolites.

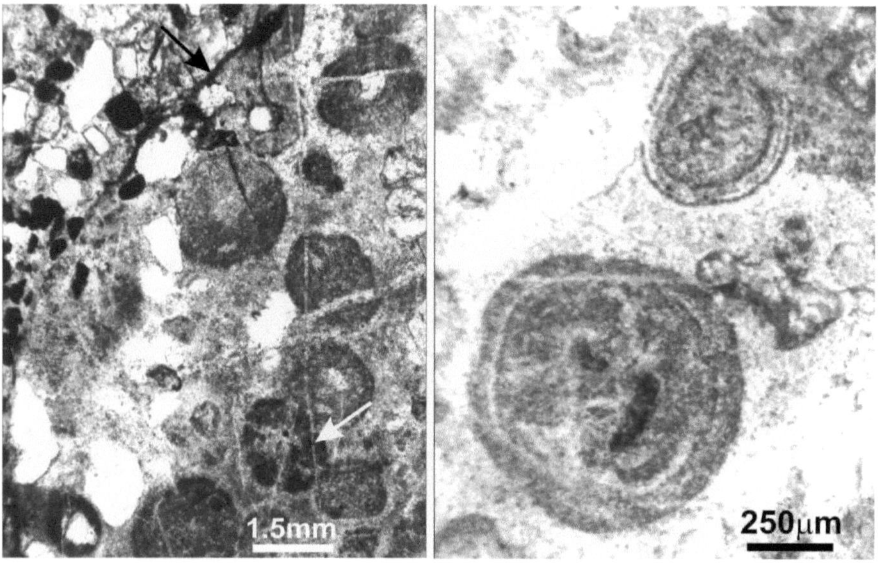


**Fig. B51** Ooid-like structures from the Strelley Pool sandstone (Photographs from the Oxford Palaeobiology Collections)

| | |
|---|---|
| Age of host rock | ~3,400 Ma: Strelley Pool sandstone, East Strelley greenstone belt, Pilbara, Western Australia; age constrained by U–Pb dates of 3,350 Ma and 3,426 Ma on zircons from overlying and underlying units respectively |
| Age of structures | At least ~3,430 Ma: Found within lithified 'intraclasts' (left image) in a conglomerate. Clasts have a discontinuous carbonaceous coating (black arrow) and occur within a black, pyrite- and carbon-rich, sandstone |
| Geological context? | Found in conglomerate deposited in beach-like setting; environment viable for life |
| Biological morphology? | Concentrically laminated round to oval structures showing close similarity to modern ooids or caliche glaebules. Some of the structures have a colloform texture (white arrow) |
| Biogeochemistry? | Laminae contain carbon |
| Type of organism? | Unknown |
| INTERPRETATION | Possible ooids (work in progress) |
| **IS THIS REALLY LIFE?** | **Undecided as yet. More geochemistry needed, and the role of biology in ooid formation, in general, needs to be better understood** |

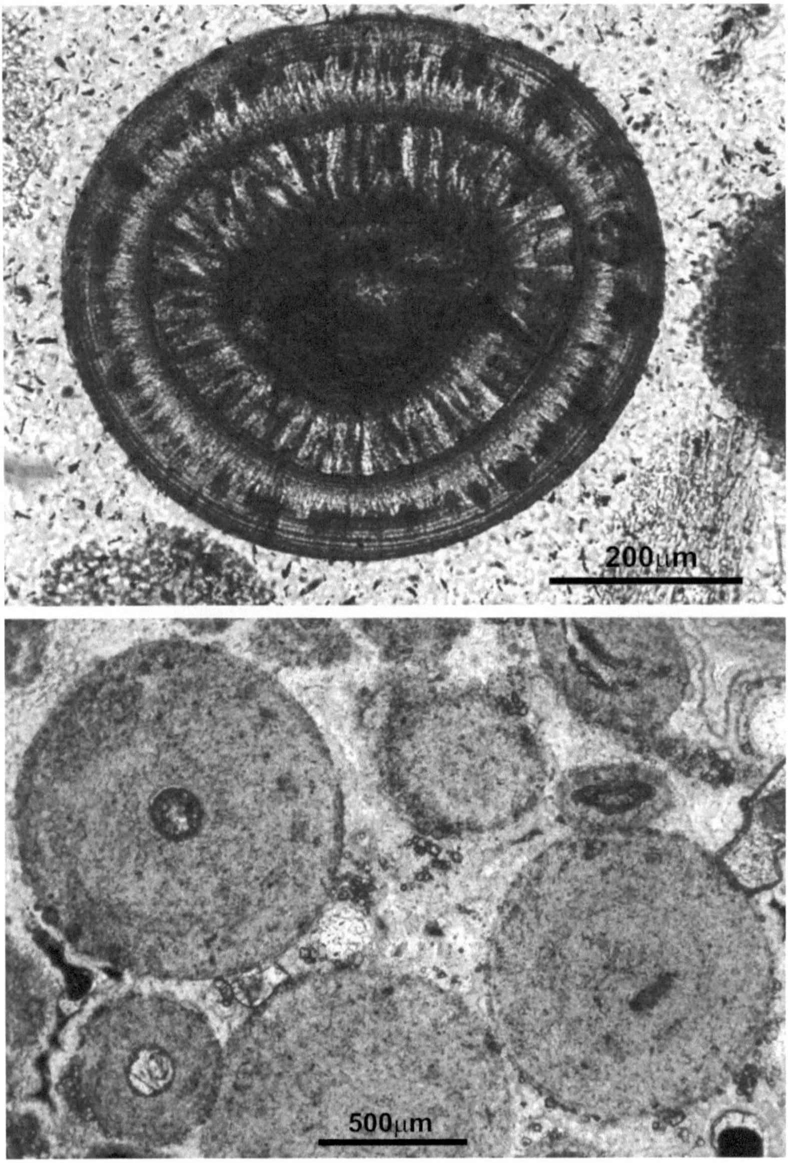


**Fig. B52** Younger analogues for the possible ooids reported from the Strelley Pool conglomerate. *Top*, recent ooid from the Great Salt Lake, Utah, USA. *Bottom*, silicified ooids from the Jurassic of England. Note how much of the concentric and radial texture seen in the upper image is lost during silicification (*lower* image) (Samples from the Oxford Palaeobiology Collections)

# Recommended Reading

Allwood, A. C., Walter, M. R., Kamber, B. S., Marshall, C. P., and Burch, I. W., 2006, Stromatolite reef from the Early Archaean era of Australia, *Nature* **441**: 714–718.

Allwood, A. C., Walter, M. R., Burch, I. W., and Kamber, B. S., 2007, 3.43 billion-year-old stromatolite reef from the Pilbara Craton of Western Australia: ecosystem-scale insights to early life on Earth, *Precambrian Research* **158**: 198–227.

Ascaso, C., and Wierzchos, J., 2002, New approaches to the study of Antarctic lithobiontic microorganisms and their inorganic traces, and their application in the detection of life in Martian rocks, *International Microbiology* **5**: 215–222.

Brasier, M. D., McLoughlin, N., and Wacey, D., 2006, A fresh look at the fossil evidence for early Archaean cellular life, *Philosophical Transactions of the Royal Society B* **361**: 887–902.

Etzel, K, Huber, H., Rachel, R., Schmalz, G., Thomm, M., and Depmeier, W., 2007, Pyrite surface alteration of synthetic single crystals as effect of microbial activity and crystallographic orientation, *Advanced Materials Research* **20–21**: 350–353.

Friedmann, E. I., and Weed, R., 1987, Microbial trace-fossil formation, biogenous, and abiotic weathering in the Antarctic cold desert, *Science* **236**: 703–705.

Hofmann, H. J., 2000, Archean stromatolites as microbial archives. In: Riding, R. E., and Awramik, S. M. (Eds.) *Microbial Sediments*, Springer, Berlin, pp. 315–327.

Hofmann, H. J., Grey, K., Hickman, A. H., and Thorpe, R.I., 1999, Origin of 3.45 Ga Coniform Stromatolites in the Warrawoona Group, Western Australia, *Bulletin of the Geological Society of America* **111**: 1256–1262.

Lowe, D. R., 1980, Stromatolites 3,400-Myr old from the Archean of Western Australia, *Nature* **284**: 441–443.

Lowe, D. R., 1983, Restricted shallow-water sedimentation of early Archaean stromatolitic and evaporitic strata of the Strelley Pool chert, Pilbara block, Western Australia, *Precambrian Research* **19**: 239–283.

Lowe, D. R., 1994, Abiological origin of described stromatolites older than 3.2 Ga, *Geology* **22**: 387–390.

Perri, E., and Tucker, M., 2007, Bacterial fossils and microbial dolomite in Triassic stromatolites, *Geology* **35**: 207–210.

Wacey, D., and Kilburn, M. R., 2008, Microbially mediated pyrite from the > 3400 Ma Strelley Pool sandstone, Western Australia, (in prep.).

Wacey, D., Kilburn, M. R., McLoughlin, N., Parnell, J., Stoakes, C. A., and Brasier, M. D., 2008a, Use of NanoSIMS to investigate early life on Earth: ambient inclusion trails in a *c.* 3400 Ma sandstone, *Journal of the Geological Society of London* **165**: 43–53.

Wacey, D., Kilburn, M. R., Stoakes, C. A., Aggleton, H., and Brasier, M.D., 2008b, Ambient inclusion trails: their recognition, age range and applicability to early life on earth. In: Dilek, Y., Furnes, H., and Muehlenbachs, K. (Eds.) *Links Between Geological Processes, Microbial Activities and Evolution of Life*, Springer, pp. 113–133.

Wacey, D., Kilburn, M., Brasier, M. D., Parnell, J., and Green, O. R., 2008c, Microbial oxidation of > 3400 Ma pyrite grains, (in prep.).

# Chapter 13
# ~3,416–3,334 Ma, Kromberg Formation, Barberton, South Africa

**Fig. B53** The Kromberg Formation consists of basalt, komatiite and mafic volcaniclastic rocks, with minor black cherts and banded cherts. Putative signs of life have been reported from both basaltic (*upper* image) and chert units (*lower* image). Scale: top, penknife is 8 cm long; bottom, pen is 15 cm long (Photographs courtesy of Nicola McLoughlin)

D. Wacey, *Early Life on Earth: A Practical Guide,*

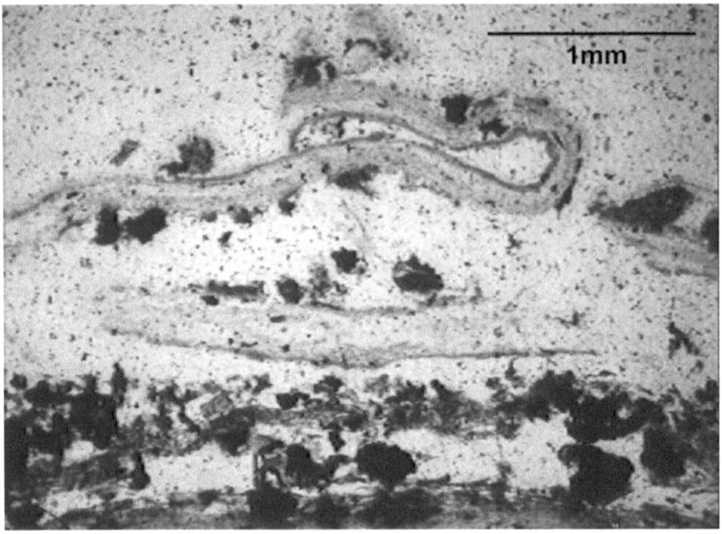


**Fig. B54** Putative microbial mat from the Kromberg Formation (Image from the Oxford Palaeobiology Collections)

| | |
|---|---|
| Age of host rock? | ~3,416 Ma: Buck Reef Chert, Kromberg Formation, Barberton Greenstone Belt, South Africa; age constrained by U–Pb date of 3,416 ± 5 Ma |
| Age of structures? | ~3,416 Ma: Primary carbonaceous material within the black layers of a black and white banded chert. Raman micro-spectroscopy is consistent with this age |
| Geological context? | Occur in a marine silicified sediment interpreted to have been deposited on a wave- and current-active shelf; viable for life |
| Biological morphology? | Fine carbonaceous laminations separated by chert laminations or lenses, locally ripped up and plastically deformed (top of image); filaments (1–1.5 µm in diameter, up to 100 µm long) are occasionally preserved within these laminations |
| Biogeochemistry? | Carbon isotopes: $\delta^{13}C$ –20‰ to –35‰ PDB |
| Type of organism? | Photosynthetic and probably anoxygenic microbes (Tice and Lowe, 2004) |
| INTERPRETATION IN THE LITERATURE | Deformed microbial mats (Tice and Lowe, 2004) |
| **IS THIS REALLY LIFE?** | **This is a well reasoned argument for life with a good understanding of the geological context. The context is vital since similar structures may occur in non-biological silica gels** |

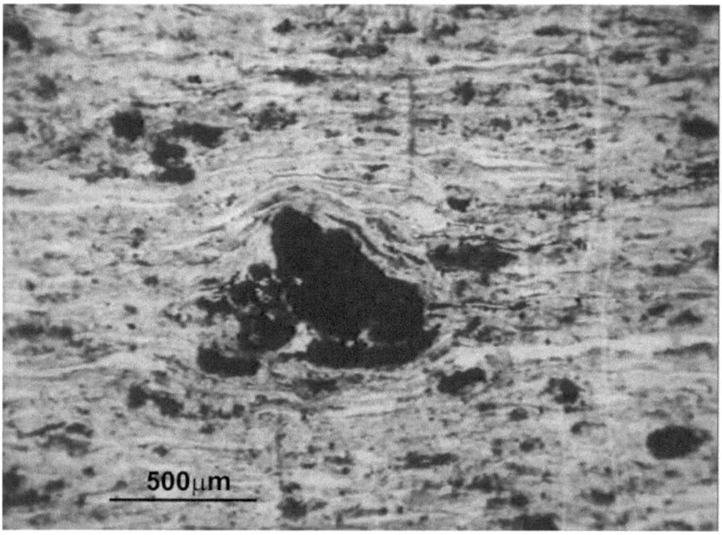


**Fig. B55** Carbonaceous grains from the Kromberg Formation (Image from the Oxford Palaeobiology Collections)

| | |
|---|---|
| Age of host rock? | ~3,416 Ma: Buck Reef Chert, Kromberg Formation, Barberton Greenstone Belt, South Africa; age constrained by U–Pb date of 3,416 ± 5 Ma on zircons in tuff inter-bedded with chert |
| Age of structures? | ~3,416 Ma: Primary carbonaceous material within the black layers of a black and white banded chert. Raman micro-spectroscopy is consistent with this age |
| Geological context? | Occur in a marine silicified sediment interpreted to have been deposited on a wave- and current-active shelf; viable for life |
| Biological morphology? | Not distinctive: Just carbonaceous laminations and carbonaceous grains; the grains are interpreted to be eroded remnants of microbial mat |
| Biogeochemistry? | Carbon isotopes: $\delta^{13}C$ −20‰ to −35‰ PDB |
| Type of organism? | Photosynthetic and probably anoxygenic microbes (Tice and Lowe, 2004) |
| INTERPRETATION IN THE LITERATURE | Eroded microbial mats (Tice and Lowe, 2004) |
| **IS THIS REALLY LIFE?** | **Less convincing than the plastically deformed structures in Fig. B54, although they do occur in the same chert sample** |

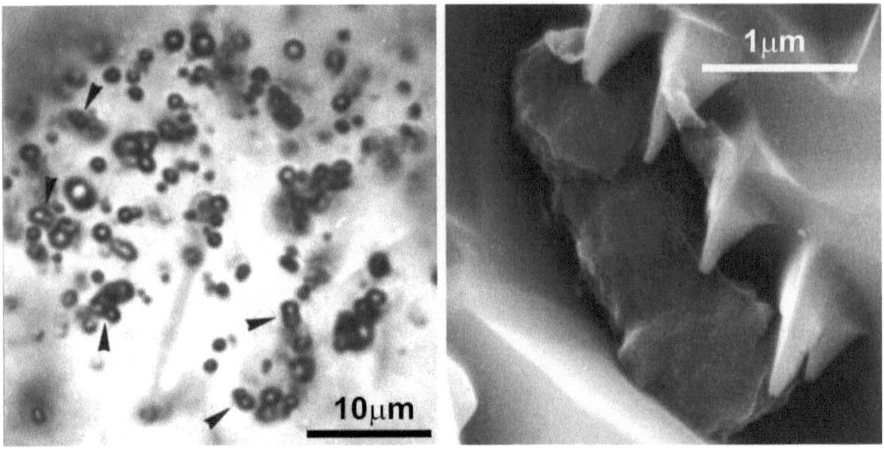


**Fig. B56** Putative microfossils from the Kromberg Formation (Images from Westall et al., 2001, reproduced with permission from Elsevier)

| | |
|---|---|
| Age of host rock? | ~3,400 Ma: Kromberg Formation, Barberton, South Africa; age constrained by U–Pb dates on zircons in tuffs at the base (3,416 ± 5 Ma) and the top (3,334 ± 3 Ma) of the formation |
| Age of structures? | ~3,400 Ma: Occur in irregular clots parallel to laminations in a millimeter-scale laminated chert. Seen in thin section (left) as well as in the SEM (right), although whether these are the same structures is debatable |
| Geological context? | Occur in bedded chert interpreted to have been deposited in shallow marine to sub-aerial setting – viable for life but not all workers agree on the interpreted environmental setting |
| Biological morphology? | Spherules: Around 1 μm in diameter with uniform size range; occur in clusters, sometimes joined in pairs (arrows in left image), zig-zags or linear strings (centre of right image); surfaces can be rough and wrinkled; spherules can also outline larger crystals |
| Biogeochemistry? | Only one bulk $\delta^{13}C$ value of −27‰ PDB; may not actually be related to these inferred bacteria |
| Type of organism? | Coccoid bacteria (Westall et al., 2001) |
| INTERPRETATION IN THE LITERATURE | Microfossils (Westall et al., 2001; Muir and Grant, 1976) |
| **IS THIS REALLY LIFE?** | **Reliant on morphology which is too simple to be attributed uniquely to microbes** |

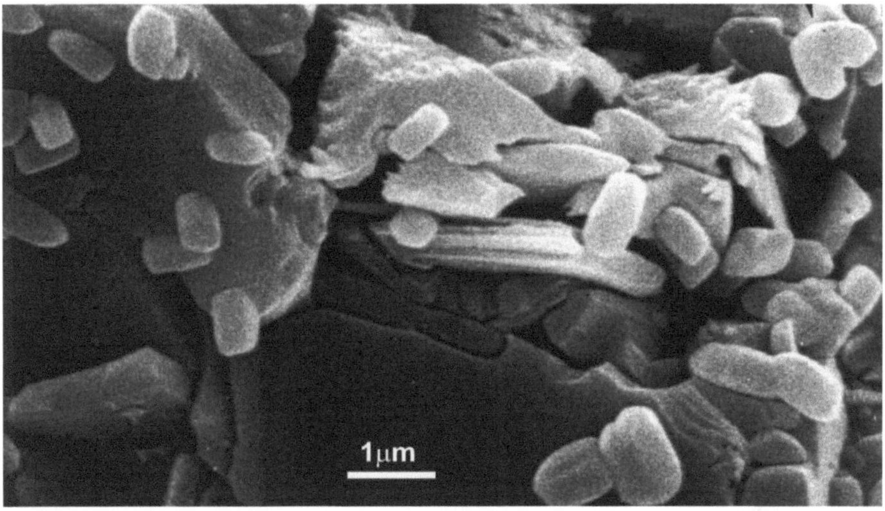


**Fig. B57** Putative rod-like microfossils from the Kromberg Formation (Image from Westall et al., 2001, reproduced with permission from Elsevier)

| | |
|---|---|
| Age of host rock? | ~3,400 Ma: Kromberg Formation, Barberton, South Africa; age constrained by U–Pb dates on zircons in tuffs at the base (3,416 ± 5 Ma) and the top (3,334 ± 3 Ma) of the formation |
| Age of structures? | ~3,400 Ma: Appear to be imbedded in the quartz matrix of primary bedded cherts |
| Geological context? | Occur in bedded chert interpreted to have been deposited in shallow marine to sub-aerial setting – viable for life but some argument over the interpreted environmental setting |
| Biological morphology? | Rod-shaped: 0.65–2 μm in length; often clustered; some joined at their extremities, (lower right of figure) others cemented together (right of scale bar in figure) |
| Biogeochemistry? | Only one bulk $\delta^{13}C$ value of –27‰ PDB; may not actually be related to these structures |
| Type of organism? | Unknown, possibly not applicable |
| INTERPRETATION IN THE LITERATURE | Possible microfossils (Westall et al., 2001) |
| **IS THIS REALLY LIFE?** | **Unlikely. Many of the proposed 'microfossils' have crystalline terminations. Formation as tiny mineral inclusions is more likely** |

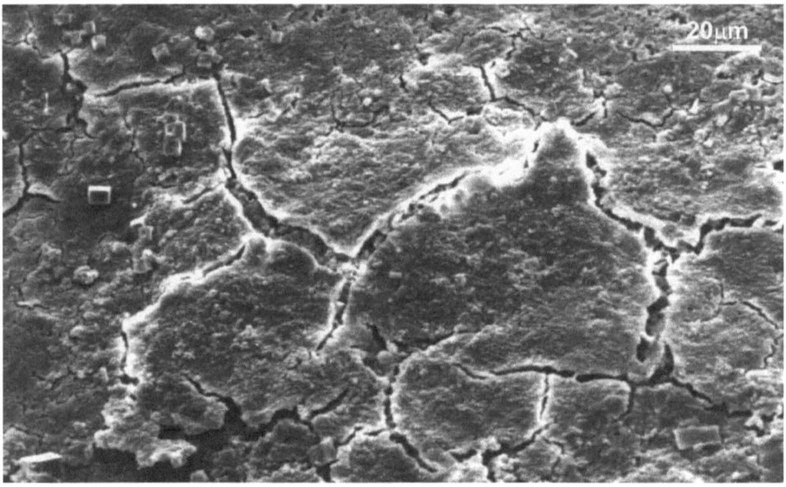


**Fig. B58** Putative microbial mat within the Kromberg Formation (Image from Westall et al., 2001, reproduced with permission from Elsevier)

| | |
|---|---|
| Age of host rock? | ~3,400 Ma: Kromberg Formation, Barberton, South Africa; age constrained by U–Pb dates on zircons in tuffs at the base (3,416 ± 5 Ma) and the top (3,334 ± 3 Ma) of the formation |
| Age of structures? | ~3,400 Ma: Granular film on a bedding plane within a sedimentary bedded chert |
| Geological context? | Bedded chert interpreted to have been deposited in shallow marine to sub-aerial setting; viable for life but not all agree on the interpreted environmental setting |
| Biological morphology? | Non-rigid laminated structures resembling modern microbially-produced biofilms; possible gas-escape structures also evident |
| Biogeochemistry? | Only one bulk $\delta^{13}C$ value of –27‰ PDB; may not be directly related to these structures |
| Type of organism? | Some unknown mat-building community |
| INTERPRETATION IN THE LITERATURE | Microbial mat (Westall et al., 2001) |
| **IS THIS REALLY LIFE?** | **Possible life, but more *in situ* geochemistry is needed to discount a purely chemical formation mechanism** |

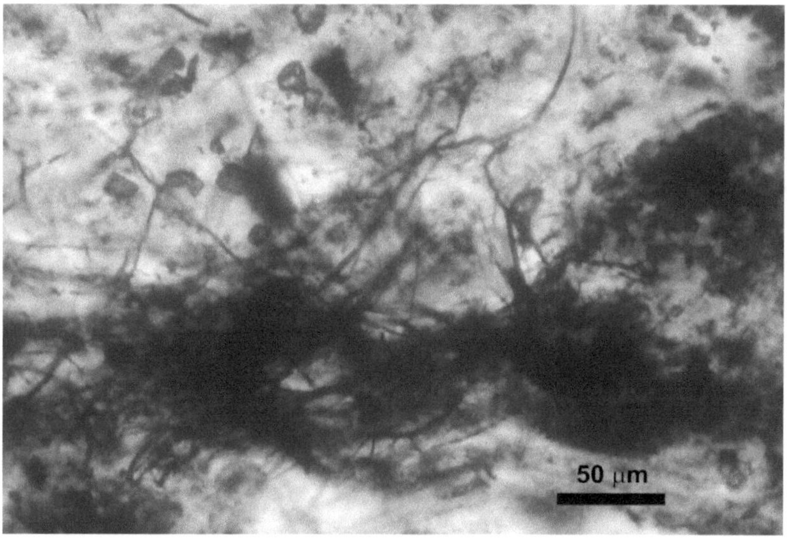


**Fig. B59** Clumps of filamentous microfossils from the Kromberg Formation (Image courtesy of Maud Walsh)

| | |
|---|---|
| Age of host rock? | ~3,400 Ma: Kromberg Formation, Barberton Greenstone Belt, South Africa; age constrained by U–Pb dates on zircons in tuffs at the base (3,416 ± 5 Ma) and the top (3,334 ± 3 Ma) of the formation |
| Age of structures? | ~3,400 Ma: Often sub-parallel to bedding or radiating away from laminated carbonaceous material; well away from veins and weathering surfaces |
| Geological context? | Occur in carbonaceous chert formed during a break in volcanic activity; habitat viable for life |
| Biological morphology? | 1. Tubular filaments: 1.5–2.5 μm in diameter, 10–150 μm in length with slight constrictions<br>2. Solid filaments: 0.5–2.5 μm in diameter, up to 200 μm long |
| Biogeochemistry? | Carbonaceous and occasionally pyritic |
| Type of organism? | Filamentous cyanobacteria (inferred by Walsh, 1992) |
| INTERPRETATION IN THE LITERATURE | Possible microfossils (Walsh and Lowe, 1985; Walsh, 1992; Brooks et al., 1973) |
| **IS THIS REALLY LIFE?** | **Possible life, but geochemical evidence is lacking; the solid examples could be simple mineral filaments** |

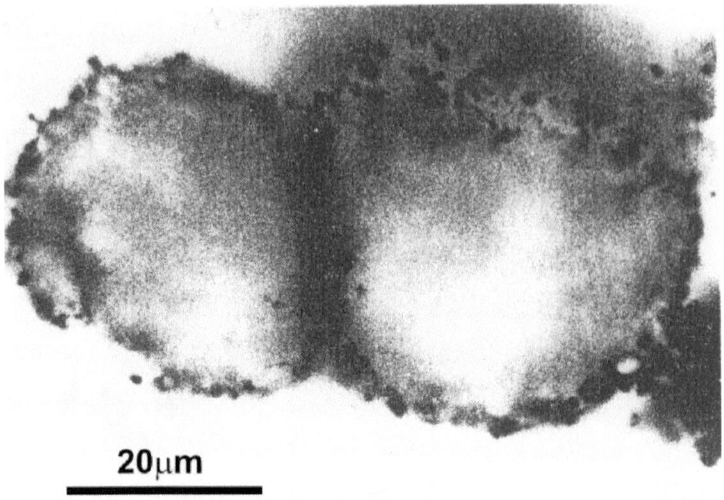


**Fig. B60** Putative spherical microfossils from the Kromberg Formation (Image from Walsh, 1992, reproduced with permission from Elsevier)

| | |
|---|---|
| Age of host rock? | ~3,400 Ma: Kromberg Formation, Barberton, South Africa; age constrained by U–Pb dates on zircons in tuffs at the base (3,416 ± 5 Ma) and the top (3,334 ± 3 Ma) of the formation |
| Age of structures? | ~3,400 Ma: Associated with primary carbonaceous laminations, well away from veins or weathered surfaces |
| Geological context? | Found in carbonaceous stratiform chert formed during break in volcanic activity; context viable for life |
| Biological morphology? | Large spheres and ellipsoids: 10–84 μm long and 10–60 μm wide; walls 2–15 μm thick; often paired (pictured above) |
| | Small spheres: 4.5–12.8 μm in diameter with 0.5 μm walls; sometimes occur in chains and clusters |
| Biogeochemistry? | Granular carbon and occasionally pyrite |
| Type of organism? | Coccoid cyanobacteria (inferred for small spheres) and sheaths enclosing cyanobacterial colonies (inferred for the large spheres) |
| INTERPRETATION IN THE LITEATURE | Small spheres interpreted as microfossils (Walsh, 1992; Brooks et al., 1973) |
| | Large spheres interpreted as sheaths, spores (Walsh, 1992) or fossilised gas bubbles (Westall et al., 2001) |
| **IS THIS REALLY LIFE?** | **Probably not life. Simple spheres are one of the most common types of non-biological artefacts and some have slightly rhombic cross sections** |

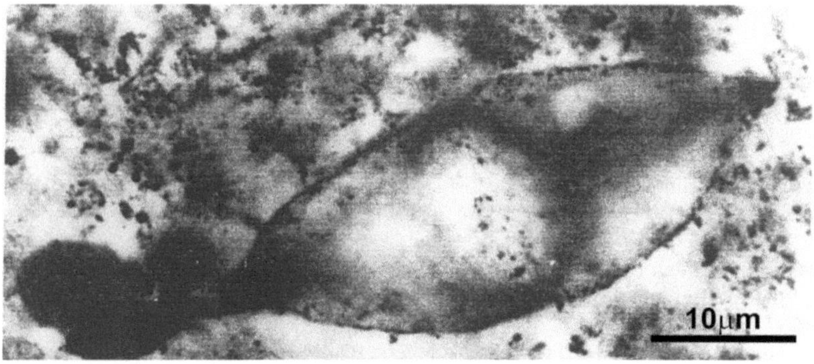


**Fig. B61** Thin section image of a carbonaceous spindle shaped structure from the Kromberg Formation (Image from Walsh, 1992, reproduced with permission from Elsevier)

| | |
|---|---|
| Age of host rock? | ~3,400 Ma: Kromberg Formation, Barberton, South Africa; age constrained by U–Pb dates on zircons in tuffs at the base (3,416 ± 5 Ma) and the top (3,334 ± 3 Ma) of the formation |
| Age of structures? | ~3,400 Ma: Found in association with primary carbonaceous laminations, away from fractures, veins and weathered surfaces |
| Geological context? | Occur in sedimentary carbonaceous chert formed during break in volcanic activity; habitat viable for life |
| Biological morphology? | Spindle shaped: Envelopes 13–135 μm long; some have one or more hollow centres |
| Biogeochemistry? | Carbonaceous and occasionally pyritic |
| Type of organism? | Unknown, possibly not applicable |
| INTERPRETATION IN THE LITERATURE | 1. Sheaths of colonies of bacterial cells *or* carbonaceous coatings of gypsum crystals (Walsh, 1992)<br>2. Bacterial gas escape structures (Westall et al., 2001) |
| **IS THIS REALLY LIFE?** | **Open to debate because these structures are poorly understood – at present, biological and non-biological explanations are both possible** |

## Recommended Reading

Brooks, J., Muir, M. D., and Shaw, G., 1973, Chemistry and morphology of Precambrian micro-organisms, *Nature* **244**: 215–217.

Byerly, G. R., Kroner, A., Lowe, D. L., Todt, W., and Walsh, M. M., 1996, Prolonged magmatism and time constraints for sediment deposition in the early Archean Barberton greenstone belt: evidence from the Upper Onverwacht and Fig Tree groups, *Precambrian Research* **78**: 125–138.

Muir, M. D., and Grant, P. R., 1976, Micropalaeontological evidence from the Onverwacht Group, South Africa. In: Windley, B. F. (Ed.) *The Early History of the Earth*, Wiley Interscience, London, pp. 595–608.

Tice, M. M., and Lowe, D. R., 2004, Photosynthetic microbial mats in the 3,416-Myr-old ocean, *Nature* **431**: 549–552.

Van Zuilen, M. A., Chaussidon, M., Rollion-Bard, C., and Marty, B., 2007, Carbonaceous cherts of the Barberton Greenstone Belt, South Africa: isotopic, chemical and structural characteristics of individual microstructures, *Geochimica et Cosmochimica Acta* **71**: 655–669.

Walsh, M. M., 1992, Microfossils and possible microfossils from the early Archean Onverwacht Group, Barberton Mountain Land, South Africa, *Precambrian Research* **54**: 271–293.

Walsh, M. M., and Lowe, D. L., 1985, Filamentous microfossils from the 3,500 Myr-old Onverwacht Group, Barberton Mountain Land, South Africa, *Nature* **314**: 530–532.

Walsh, M. M., and Lowe, D. L., 1999, Modes of accumulation of carbonaceous matter in the early Archean: a petrographic and geochemical study of the carbonaceous cherts of the Swaziland Supergroup In: Lowe, D. R., and Byerley, G. R. (Eds.) *Geologic Evolution of the Barberton Greenstone Belt, South Africa, GSA Special Paper* **329**, Boulder, CO, pp. 167–188.

Westall, F., de Witt, M. J., Dann, J., van der Gaast, S., de Ronde, C. E. J., and Gerneke, D., 2001, Early Archean fossil bacteria and biofilms in hydrothermally-influenced sediments from the Barberton greenstone belt, South Africa, *Precambrian Research* **106**: 93–116.

# Chapter 14
# ~3,350 Ma, Euro Basalt, East Pilbara, Western Australia

**Fig. B62** The Euro Basalt is a 5–8 km thick unit, made up of komatiitic, tholeiitic and high magnesium basalt, often pillowed (*upper* image), with occasional thin (<10 m) chert, fined grained siliciclastic and volcaniclastic horizons (*lower* image). Pillows are well preserved in many areas and putative evidence of life occurs in the dark outermost margins of these pillows and in glass shards preserved in inter-pillow hyaloclastite. Scale: top, hammer is 40 cm long; bottom, sunglasses are 15 cm long (Photographs by the author)

D. Wacey, *Early Life on Earth: A Practical Guide*,

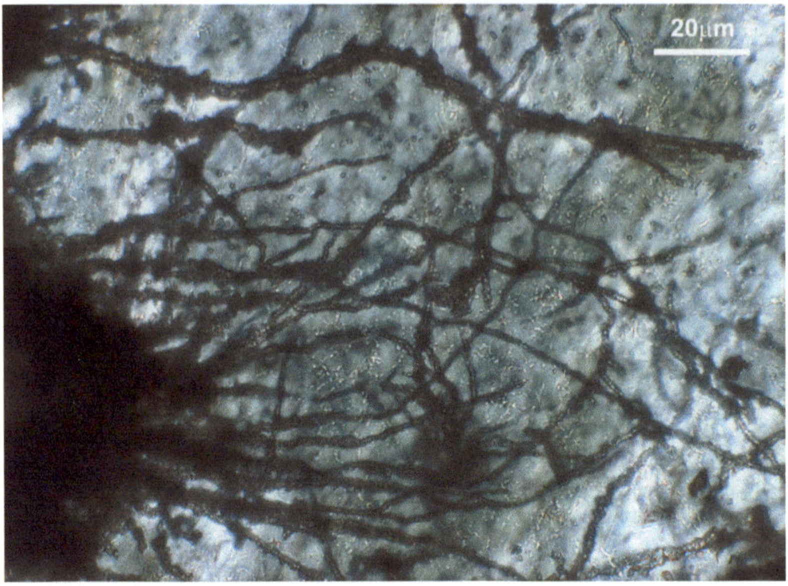


**Fig. B63** Thin section image of micro-tubular structures within the Euro Basalt (Photograph courtesy of Nicola McLoughlin)

| | |
|---|---|
| Age of host rock? | ~3,350 Ma, Euro Basalt, Pilbara, Western Australia; age constraint is from U–Pb dating of zircon from a felsic unit towards the base of the Euro Basalt |
| Age of structures? | Between 3,350 Ma and ~2,900 Ma from U–Pb dating of a titanite mineral phase infilling the microtubes |
| Geological context? | Found around re-healed fractures in the formerly glassy margins of pillow basalts and in inter-pillow hyaloclastite (glass shards that have spalled off the pillows) |
| Biological morphology? | Tubular: 1–5 μm in width and up to 150 μm in length; often branched and segmented and show sharp changes in direction; show similarity to microbially mediated microtubes in modern basalts |
| Biogeochemistry? | Possible enrichment of carbon and nitrogen within tubes (but spatial resolution of microprobe technique is poor) |
| Type of organism? | Unknown, possibly thermophilic endolith |
| INTERPRETATION IN THE LITERATURE | Trace fossil (Banerjee et al., 2007) |
| **IS THIS REALLY LIFE?** | **Possible life, but the formation of even the modern examples is still poorly understood. Higher resolution geochemistry also needed** |

**Fig. B64** Younger analogues for the putative trace fossils from the Euro Basalt pillow lavas. *Top*, biologically mediated micro-tubular structures from volcanic glass (122 Ma Ontong Java Plateau). *Bottom*, a variety of micro-tubular structures (five examples *arrowed*) in glass shards within inter-pillow hyaloclastite (92 Ma Troodos Ophiolite) (Images courtesy of Nicola McLoughlin)

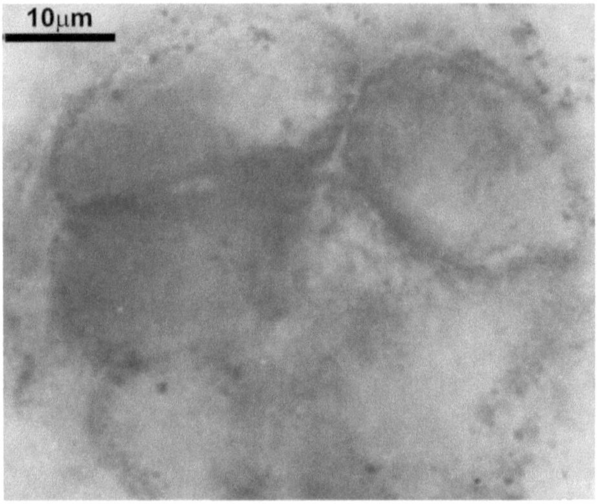


**Fig. B65** Thin section image of putative microfossils, possibly from a chart unit within the Euro Basalt (Image from Schopf and Packer, 1987, reproduced with permission from AAAS)

| | |
|---|---|
| Age of host rock? | Debatable: Originally assigned to the Towers Fm (name no longer in use), but grid reference in original report places it within a thin chert of the ~3,350 Ma Euro Basalt |
| Age of structures? | ~3,350 Ma, if correctly located: Occur in thin sections of a bedded sediment of known age; occur away from later cross cutting chert veins |
| Geological context? | Occur in finely laminated organic rich zones of a black and white stratiform chert deposited during a pause in volcanic activity; viable for life |
| Biological morphology? | 2 types of 'colonies'. One containing spheroidal 'cells' of ~8 μm diameter. The other (pictured above) containing 'cells' of ~21 μm diameter sometimes enclosed by a 'sheath' |
| Biogeochemistry? | None |
| Type of organism? | Cyanobacteria (chroococcales) (Schopf and Packer, 1987) |
| INTERPRETATION IN THE LITERATURE | Sheath enclosed colonial unicells (i.e. microfossils) (Schopf and Packer, 1987) |
| **IS THIS REALLY LIFE?** | **Possible life, but debatable location and lack of geochemical evidence count against it. Original identification as cyanobacteria is very controversial** |

# Recommended Reading

Alt, J. C., and Mata, P., 2000, On the role of microbes in the alteration of submarine basaltic glass: a TEM study, *Earth and Planetary Science Letters* **181**: 301–313.

Banerjee, N. R., and Muehlenbachs K., 2003, Tuff life: bioalteration in volcaniclastic rocks from the Ontong Java Plateau, *Geochemistry Geophysics Geosystems* **4**: 2002GC000470.

Banerjee, N. R., Simonetti, A., Furnes, H., Muehlenbachs, K., Staudigel, H., Heaman, L., and Van Kranendonk, M. J., 2007, Direct dating of microbial ichnofossils, *Geology* **35**: 487–490.

Fisk, M. R., Storrie-Lombardi, M. C., Douglas, S., Popa, R., McDonald, G., and Di Meo-Savoie, C., 2003, Evidence of biological activity in Hawaiian subsurface basalts, *Geochemistry Geophysics Geosystems* **4**: 2003GC000387.

Furnes, H., Staudigel, H., Thorseth, I. H., Torsvik, T., Muehlenbachs, K., and Tumyr, O., 2001, Bioalteration of basaltic glass in the oceanic crust, *Geochemistry Geophysics Geosystems* **2**: 2000GC000150.

Schopf, J. W., and Packer, B. M., 1987, Early Archean (3.3 Billion to 3.5 Billion-Year-Old) Microfossils from Warrawoona Group, Australia, *Science* **237**: 70–73.

Staudigel, H., Chastain, R. A., Yayanos, A., and Bourcier, R., 1995, Biologically mediated dissolution of glass, *Chemical Geology* **126**: 119–135.

Staudigel, H., Furnes, H., Banerjee, N. R., Dilek, Y., and Muehlenbachs, K., 2006, Microbes and volcanoes: a tale from the oceans, ophiolites and greenstone belts, *GSA Today* **16**: 4–11.

Thorseth, I. H., Furnes, H., and Heldal, M., 1992, The importance of microbiological activity in the alteration of natural basaltic glass, *Geochimica et Cosmochimica Acta* **56**: 845–850.

# Chapter 15
# ~3,250 Ma, Fig Tree Group, Barberton, South Africa

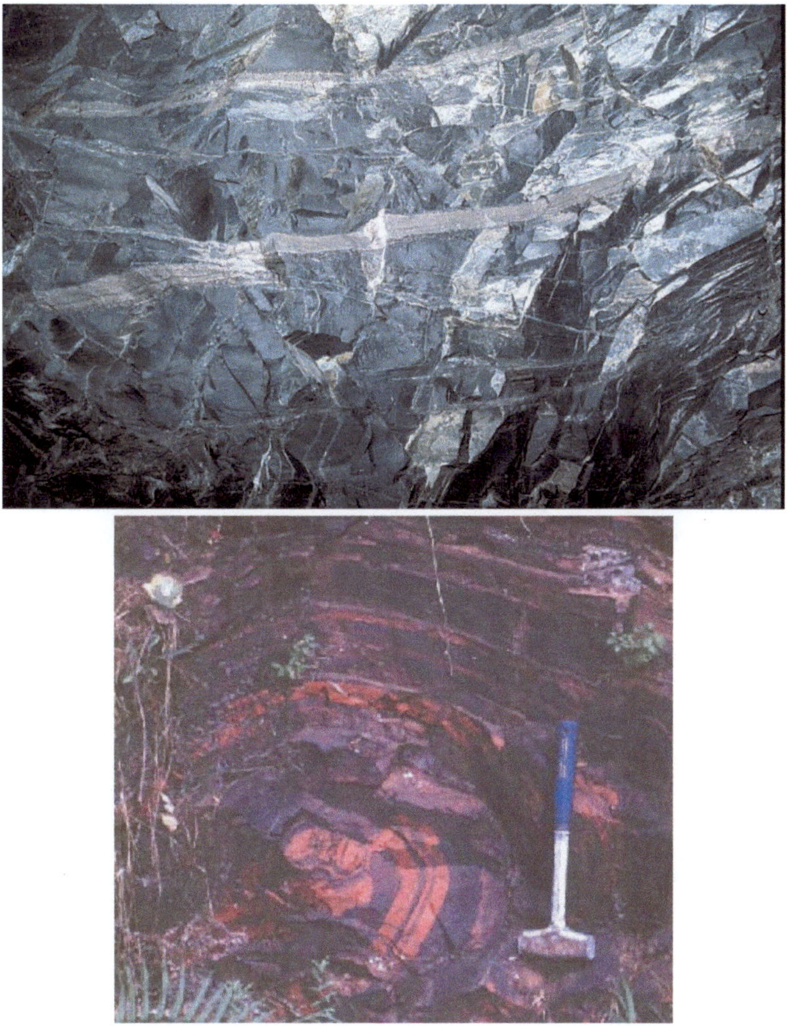

**Fig. B66** The Fig Tree Group consists of terrigenous clastic sedimentary units interstratified with dacitic to rhyodacitic volcaniclastic and volcanic rocks. The sedimentary units comprise a thick sequence of graywackes, slates, and shales (*upper* image) with interbedded horizons of chert, jasper, and ironstone (*lower* image). Putative evidence for life has been reported from carbonaceous chert horizons within this Group. Scale: hammer is 60 cm long, both images at same scale

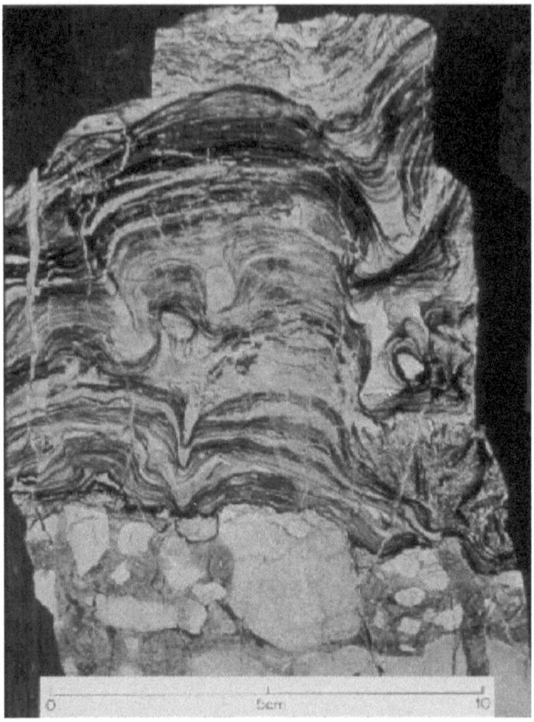


**Fig. B67** Polished slab of a putative stromatolite from the Fig Tree Group (Photograph courtesy of Gary Byerly)

| | |
|---|---|
| Age of host rock? | ~3,245 Ma: Sheba Formation, Fig Tree Group, Barberton, South Africa; age from U–Pb dating of zircons within the formation |
| Age of structures? | ~3,245 Ma: Indigenous (bio?)sedimentary structure within stratiform chert |
| Geological context? | Occur as part of a sedimentary sequence of ferruginous and carbonaceous cherts, volcanic ash and sandstones. Accumulated in quiescent periods between komatiitic lava flows. Habitat viable for life |
| Biological morphology? | Laterally linked domes, 1–3 cm wide and 0.5–3 cm high; pseudo-columns up to 10 cm high, plus crinkly stratiform structures. Synoptic relief on laminae up to 1.5 mm; inheritance of laminae shapes |
| Biogeochemistry? | None |
| Type of organism? | Inferred filamentous and/or coccoid mat-building bacteria or cyanobacteria (Byerly et al., 1986) |
| INTERPRETATION IN THE LITERATURE | Stromatolite (Byerly et al., 1986) |
| **IS THIS REALLY LIFE?** | **Possible life, more evidence needed. Morphological complexity similar to modern stromatolites, but non-biological experiments produce similar morphologies** |

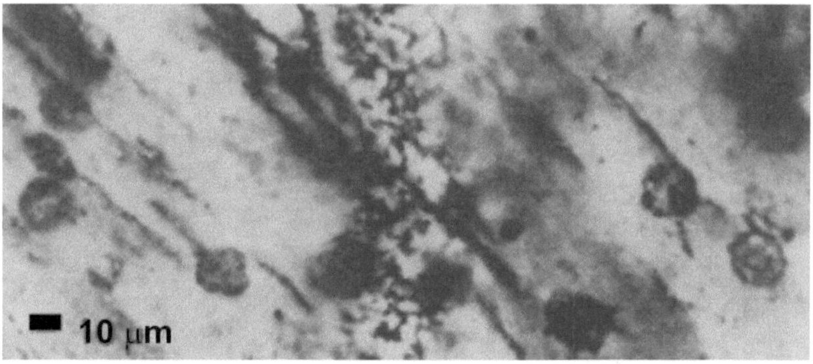


**Fig. B68** Thin section image of putative microfossils (roughly spherical objects) from the Fig Tree Group (Image from Schopf and Barghoorn, 1967, reproduced with permission from AAAS)

| | |
|---|---|
| Age of host rock? | ~3,245 Ma: Sheba Formation, Fig Tree Group, Barberton, South Africa; age constrained by U–Pb dating of zircons |
| Age of structures? | ~3,245 Ma: Occur in black cherts interpreted as primary in origin; carbonaceous material is indigenous to the chert, mostly aligned parallel to bedding planes |
| Geological context? | None given |
| Biological morphology? | Spheroidal (some distorted and flattened): Limited size range, average 18.7 µm diameter; reticulate surface texture; wall ~1 µm thick; occasionally internal organic matter may be preserved within the spheres |
| Biogeochemistry? | Carbonaceous composition. No other geochemistry performed on these samples |
| Type of organism? | Unicellular and alga-like (Schopf and Barghoorn, 1967) |
| INTERPRETATION IN THE LITERATURE | Possible microfossils (Schopf and Barghoorn, 1967) Fossilised biogenic gas bubbles (Westall et al., 2001) |
| **IS THIS REALLY LIFE?** | **No geological mapping or geochemistry reported; very simple morphology so just as likely to be non-biological** |

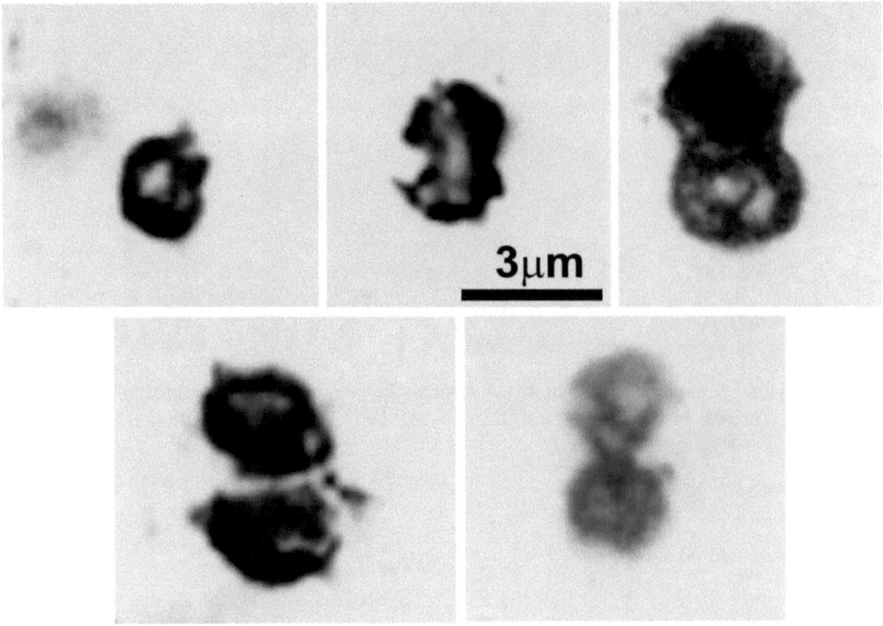


**Fig. B69** Putative spherical microfossils from the Barberton greenstone belt (Image from Knoll and Barghoorn, 1977, reproduced with permission from AAAS)

| | |
|---|---|
| Age of host rock? | ~3,260 Ma: Swartkoppie Formation (name no longer in use), around the contact of Onverwacht and Fig Tree Groups, Barberton, South Africa |
| Age of structures? | ~3,260 Ma: Microstructures appear indigenous to the chert, found only along organic rich bedding planes |
| Geological context? | Found in a carbonaceous bedded chert, deposited in a near shore environment. Environment viable for life |
| Biological morphology? | Spherical: 1–4 μm in diameter; often flattened, wrinkled or folded; narrow unimodal size frequency distribution; occasionally have internal organic contents; show stages of binary division (pictured above) |
| Biogeochemistry? | Organic walled. No other geochemistry reported |
| Type of organism? | Unknown |
| INTERPRETATION IN THE LITERATURE | Possible microfossils (Knoll and Barghoorn, 1977) |
| **IS THIS REALLY LIFE?** | **Possible life. Simple morphology and absence of any biogeochemical tests means that these structures cannot be unambiguously identified as biological** |

# Recommended Reading


Byerly, G. R., Lowe, D. L., and Walsh, M. M., 1986, Stromatolites from the 3,300–3,500-Myr Swaziland Supergroup, Barberton Mountain Land, South Africa, *Nature* **319**: 489–491.

Byerly, G. R., Kroner, A., Lowe, D. L., Todt, W., and Walsh, M. M., 1996, Prolonged magmatism and time constraints for sediment deposition in the early Archean Barberton greenstone belt: evidence from the Upper Onverwacht and Fig Tree groups, *Precambrian Research* **78**: 125–138.

Knoll, A. H., and Barghoorn, E. S., 1977, Archean microfossils showing cell division from the Swaziland System of South Africa, *Science* **198**: 396–398.

Schopf, J. W., and Barghoorn, E. S., 1967, Alga-like fossils from the early Precambrian of South Africa, *Science* **156**: 508–512.

Westall, F. de Witt, M. J., Dann, J., van der Gaast, S., de Ronde, C. E. J., and Gerneke, D., 2001, Early Archean fossil bacteria and biofilms in hydrothermally-influenced sediments from the Barberton greenstone belt, South Africa, *Precambrian Research* **106**: 93–116.

# Chapter 16
# ~3,240 Ma, Kangaroo Caves Formation, East Pilbara, Western Australia

**Fig. B70** The Kangaroo Caves Formation is notable for containing one of Earth's oldest and best preserved volcanic hosted massive sulphide deposits (VHMS). Putative biological structures have been described both from the VHMS (e.g., from the core pictured *above* where gold colour indicates pyrite) and from surrounding and overlying sediments (Photograph from the Oxford Palaeobiology Collections)

## 16.1 Summary of Claims for Early Life from this Formation

Pyritic filaments from within the VHMS deposit were first reported by Rasmussen (2000) and interpreted as microfossils, likely thermophilic, chemotrophic prokaryotes. These filaments are 0.5–2.0 μm in width and up to 300 μm long, can be straight, curved or sinuous and exhibit putative biological behaviour including preferred orientations, clustering and intertwining (Fig. B71). They only occur in early mineral phases that are clearly cross cut by later fractures.

Non-biological fibrous mineral growths are relatively common in hydrothermal ore deposits so this is the null hypothesis that needs to be rejected here. Recollection and re-examination of this material by the author shows that these filaments differ from non-biological ones in being unbranched, of constant diameter, and distinctively entangled. There is as yet, however, no evidence for cellular organisation nor for metabolic processing. Even so, this is an intriguing discovery consistent with the hypothesis of a thermophilic habitat for primitive life forms, in the vicinity of sub-marine hydrothermal vents at depths of ~1,000 m, below the light penetration zone.

Two other possible pieces of evidence for life in this Formation come from sediments found just above the VHMS deposit within a diamond drill core. These are tubular bundles of filaments and spheres reported by Duck et al. (2007) (Fig. B72 and page 82), and ambient inclusion trails reported by Wacey et al. (2008a, b) (Fig. B73, see page 183 for a discussion of the biogenicity of ambient inclusion trails).

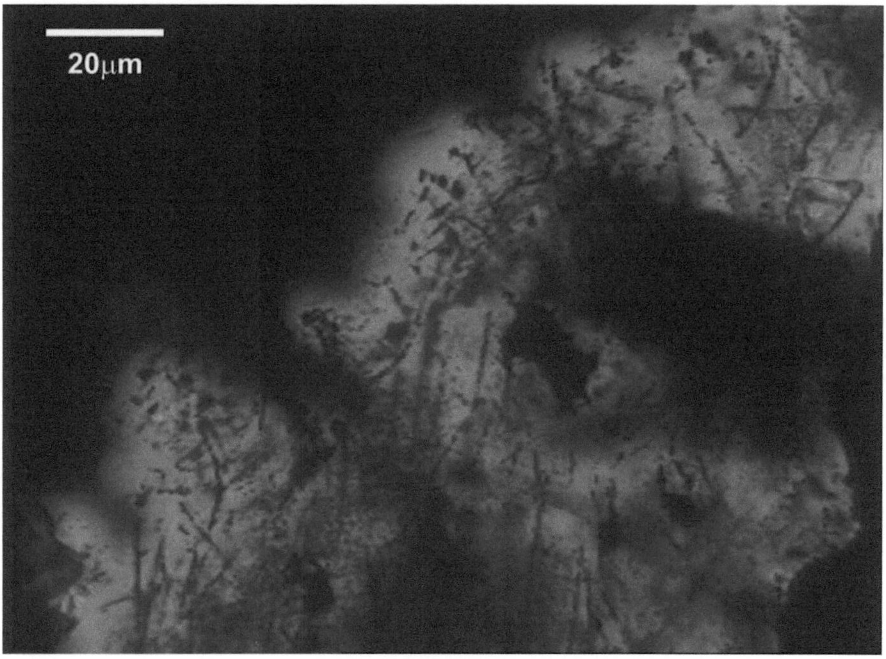


**Fig. B71** Thin section image of pyritised putative filamentous microfossils from the Kangaroo Caves Formation (Photograph from the Oxford Palaeobiology Collections)

| | |
|---|---|
| Age of host rock? | ~3,240 Ma: Massive sulphide deposit, Kangaroo Caves Formation, Pilbara; age constrained by Pb–Pb model age of 3,260 Ma from galena, and by a 3,235 Ma age from the intrusive (therefore younger) Strelley granite |
| Age of structures? | ≥3,235 Ma: They occur in paragenetically early minerals in samples from a diamond drill core. Both the host quartz and the putative microfossils were then almost entirely replaced by 3,235 Ma sulphide mineralization |
| Geological context? | Likely thermophilic mode of life near to discharging hydrothermal fluids |
| Biological morphology? | Filamentous: Sinuous, intertwined, and densely clustered; uniform thickness, average 0.5–2 μm width and up to 300 μm length; changes in preferred orientation may indicate biological behaviour |
| Biogeochemistry? | None |
| Type of organism? | Inferred anaerobic, thermophilic, chemotrophic microbes – possible sulphur based metabolism |
| INTERPRETATION IN THE LITERATURE | Possible microfossil (Rasmussen, 2000) |
| **IS THIS REALLY LIFE?** | **Possible life, morphology is very promising. Some biogeochemical evidence now needed** |

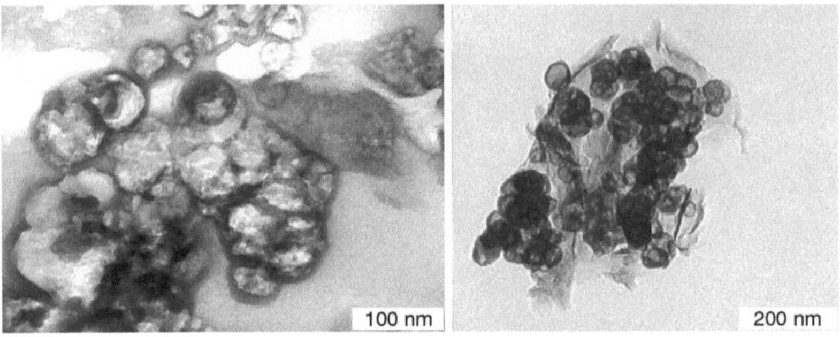


**Fig. B72** Putative microfossils from the Kangaroo Caves Formation (Images from Duck et al., 2007, courtesy of Lawrie Duck and reproduced with permission from Elsevier)

| | |
|---|---|
| Age of host rock? | ~3,240 Ma sediments immediately above the Sulphur Springs volcanic hosted massive sulphide deposit, Pilbara, Western Australia |
| Age of structures? | Appear syngenetic with ~3,240 Ma sedimentation: Organic matter parallel to bedding; eroded rafts of bedded organics within overlying breccia |
| Geological context? | Found in organic rich silicified sediments. Viable for life |
| Biological morphology? | 1. Filaments and tubes: Bundles of tubes 1–5 µm in width (cross section shown above left), up to 100 µm in length; individual tubes <1 µm wide<br>2. Spheres <50–100 nm in diameter; clusters and aggregates; possible cell budding; non-uniformity in size; these spheres (above right) may have become detached from the larger tubes |
| Biogeochemistry? | $\delta^{13}C$ bulk organics (–26.8‰ to –34‰ PDB); correlation between $^{13}C$ depletion and H/C ratio; inverse correlation between H/C ratio and thermal maturity |
| Type of organism? | Comparisons made to the tubular hyper-thermophilic achaeon *Pyrodictium abyssi* by Duck et al. (2007) |
| INTERPRETATION IN THE LITERATURE | Possible microfossils (Duck et al., 2007) |
| **IS THIS REALLY LIFE?** | **Possible life, but these structures are not seen *in situ* in the rock. The rock has been powdered so their exact age relationships cannot be determined. The very small size of the spheres (interpreted as cells) is also of concern** |

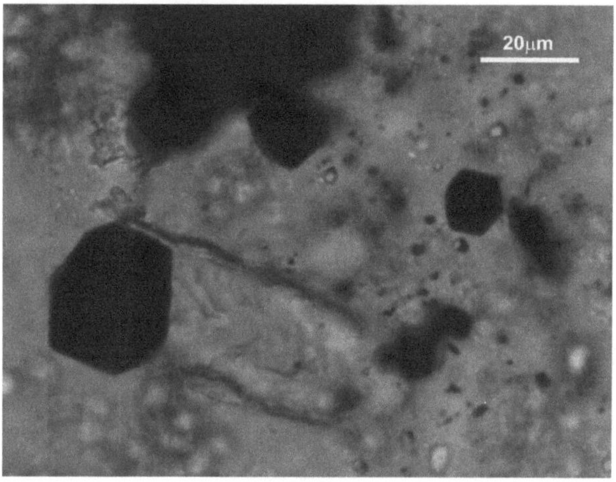


**Fig. B73** An ambient inclusion trail from the Kangaroo Caves Formation where the pyrite grain has been propelled from the centre to the left of the figure (Photograph from the Oxford Palaeobiology Collections)

| | |
|---|---|
| Age of host rock? | ~3,240 Ma Kangaroo Caves Formation, Pilbara, Western Australia; age constrained by a Pb–Pb date of 3,220 Ma on galena from the overlying Corboy Formation |
| Age of structures? | ~3,240–2,900 Ma: Movement of pyrite (black grains above) could have been any time prior to, or during, the multiple metamorphic events affecting the Pilbara in the early to mid Archean |
| Geological context? | Found in sedimentary stratiform chert. Viable for life |
| Biological morphology? | Microtubular structures: Range from 15 to 100 μm in diameter and up to ~100 μm in length; microtubes can be straight, curved, twisted or even helical, and are occasionally branched; diameter appears to remain constant along their length; commonly cluster around clots of organic matter |
| Biogeochemistry? | Similar examples in nearby rocks show carbon and nitrogen enrichments and $\delta^{13}C$ of −26‰ PDB in residual organic matter |
| Type of organism? | Unknown, likely sulphur based metabolism |
| INTERPRETATION IN THE LITERATURE | Ambient inclusion trail, possibly biologically mediated (Wacey et al., 2008a, b) |
| **IS THIS REALLY LIFE?** | **Possible life. Biogenicity only inferred from other ambient inclusion trails studied in more detail. More evidence needed from this locality** |

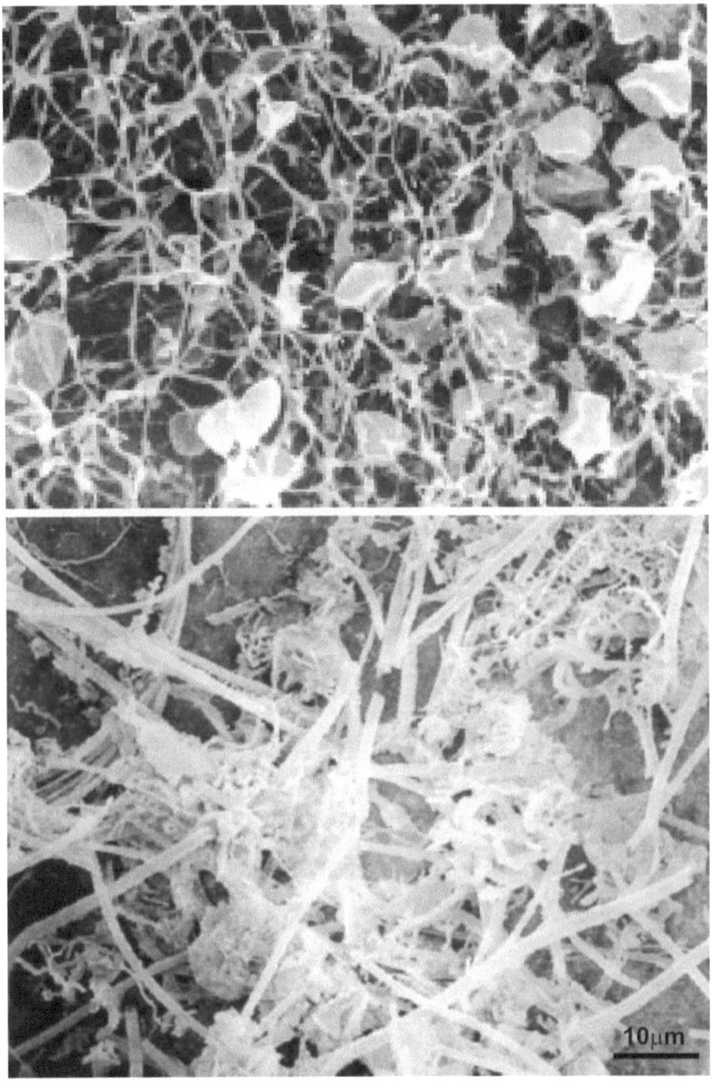


**Fig. B74** Younger analogues for putative biological structures found in the ~3,240 Ma Kangaroo Caves Formation. *Upper image* – cells and tubules of the hyper-thermophilic Archaea *Pyrodictium abyssi* (Rieger et al., 1995, reproduced with permission from Elsevier), a modern analogue for the organic remains described by Duck et al. (2007). *Lower image* – modern bacterial structures (sheaths, stalks and filaments) associated with submarine hydrothermal activity along the Juan de Fuca Ridge, possible analogues for the structures described by Rasmussen (2000) (Photograph courtesy of Grant Ferris)

## Recommended Reading

Duck, L. J., Glikson, M., Golding, S. D., and Webb, R. E., 2007, Microbial remains and other carbonaceous forms from the 3.24 Ga Sulphur Springs black smoker deposit, Western Australia, *Precambrian Research* 154: 205–220.

Kennedy, C. B., Scott, S. D., and Ferris, F. G., 2003, Characterization of bacteriogenic iron oxide deposits from Axial Volcano, Juan de Fuca Ridge, Northeast Pacific Ocean, *Geomicrobiology Journal* 20: 199–214.

Rasmussen, B., 2000, Filamentous microfossils in a 3235-million-year-old volcanogenic massive sulphide deposit, *Nature* 405: 676–679.

Rieger, G., Rachel, R., Hermann, R., and Stetter, K. O., 1995, Ultrastructure of the hyperthermophilic Archaeon *Pyrodictium abyssi*, *Journal of Structural Biology* 115: 78–87.

Vearncombe, S., Barley, M. E., Groves, D. I., McNaughton, N. J., Mikucki, E. J., and Vearncombe, J. R., 1995, 3.26 Ga black smoker type mineralization in the Strelley Belt, Pilbara Craton, Western Australia, *Journal of the Geological Society of London* 152: 587–590.

Wacey, D., Kilburn, M. R., McLoughlin, N., Parnell, J., Stoakes, C. A., and Brasier, M. D., 2008a, Use of NanoSIMS to investigate early life on Earth: ambient inclusion trails in a *c*. 3400 Ma sandstone, *Journal of the Geological Society of London* 164: 1–11.

Wacey, D., Kilburn, M. R., Stoakes, C. A., Aggleton, H., and Brasier, M. D., 2008b, Ambient inclusion trails: their recognition, age range and applicability to early life on earth. In: Dilek, Y., Furnes, H., and Muehlenbachs, K. (Eds.) *Links Between Geological Processes, Microbial Activities and Evolution of Life*. Springer, Dordrecht, The Netherlands, pp. 113–133.

# Chapter 17
# ~3,200 Ma, Moodies Group, Barberton, South Africa

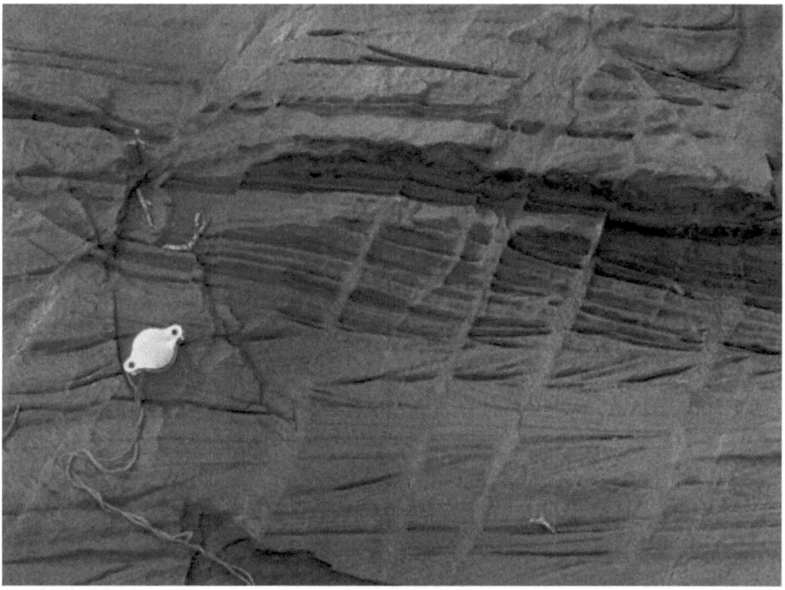

**Fig. B75** The Moodies Group consists of nearly 4 km of dominantly alluvial to shallow-marine sandstone (cross bedded sandstone pictured above), with occasional conglomerate, mudstone and volcanics, all deposited in a little over a 100 Ma time-span. Within the lower sandstone members a suite of unusual sedimentary structures have been attributed to bio-physical interactions. These include wrinkle structures, roll-up structures, laminated carbonaceous textures and desiccation cracks, interpreted to represent the remnant activity of microbial mats that stabilised these siliciclastic sediments. Comparable structures are found in similar tidal siliciclastic environments throughout the younger geological record. Scale: hand lens is 5 cm across

D. Wacey, *Early Life on Earth: A Practical Guide*,

## 17.1 Microbially Influenced Sedimentary Structures (MISS)

The Moodies Group is notable for containing a new type of potential biosignature. These are termed microbially influenced sedimentary structures (MISS). These types of structures only occur in silicilastic sediments (i.e., sandstones). They occur when the bedding in a sandstone is modified by microbial mats. Studies of modern examples show that this manifests itself in two ways (see Gerdes et al., 2000; Noffke, 2000 for a catalogue of images):

- Structures can be preserved *on* bedding planes; these include levelled depositional surfaces, wrinkles, microbial mat chips, mat curls and shrinkage cracks.
- Structures can be preserved *within* individual beds giving characteristic fabrics in thin section; these include sponge pore fabrics, gas domes, fenestrae, sinoidal laminae, and oriented grains that appear to float in a carbonaceous mat matrix.

It has been shown that for MISS to be preserved in the rock record there needs to be both initial ecological conditions that favour mat development and taphonomic conditions that allow its preservation (Noffke et al., 2002). A study of wrinkle structures from a siliciclastic shelf setting in the Neoproterozoic Nama Group of Namibia showed that mat colonization is favoured in clean, fine grained translucent quartz sands, deposited at sites where hydrodynamic flow is sufficient to sweep mud debris from mat surfaces, but insufficient to erode the bio-stabilised laminae (Noffke et al., 2002). Preservation is then achieved when the bacterial community is buried by sufficient sediment to block continued photosynthesis and prevent upward migration of cyanobacteria. Burial must of course occur without erosional destruction of the mats. Wrinkle structures are subsequently induced by further sedimentation generating loading pressure on the mats which undergo hydroplastic deformation.

These stringent ecological and taphonomic conditions mean that MISS are rarely preserved in the rock record. However, in those unique taphonomic windows where MISS are preserved, they represent a biological signature that has so far not been mimicked by physical processes alone.

The oldest described MISS come from South Africa. They have been described from the 2,900 Ma Pongola Supergroup, 2,900 Ma Witwatersrand Supergroup, and most significantly the lower part of the Moodies Group (Noffke et al., 2006b). Here, wrinkle structures (Fig. B76), desiccation cracks (Fig. B76), roll up structures and supporting carbon isotopes values of $\delta^{13}C = -21.5‰$ PDB bare testament to the former presence of microbial mats stabilizing sediment on some of Earth's earliest siliciclastic tidal flats. Care must be taken that MISS are not mistaken for tectonic features; they should occur in quartz rich sandstones of low metamorphic grade that have experienced little deformation.

Recent experiments have demonstrated a range of microbial features that can develop on silica rich substrates (e.g., Brehm et al., 2005). Microbial communities can form natural biofilms on quartz, chert or silica glass that locally elevate pH to values needed for silica dissolution (pH > 9). Alternatively, different microbial communities can produce organic acids which form complexes with silica and enhance dissolution rates (e.g., Bennett and Siegel, 1987). In this way, the microbes biochemically etch

**Fig. B76** Biosedimentary structures in the Moodies Group. *Left*, wrinkle mat and desiccation cracks on a sandstone bedding surface. *Right*, Thin section image of the fine scale scale fabric of the proposed wrinkle mat showing interwoven quartz- and carbon-rich laminae (Images courtesy of Nora Noffke and under the fair usage policy of the Geological Society of America)

| | |
|---|---|
| Age of host rock? | ~3,200 Ma: Moodies Group, Barberton greenstone belt, South Africa; maximum age constrained by geochronology on ignimbrites (3,226 ± 1 Ma), porphyries (3,222 ± 10 Ma), and dacitic clasts in conglomerates (3,225 ± 3 Ma) at the top of the underlying Fig Tree Group; minimum age is constrained by the crosscutting Salisbury Kop pluton (3,079 ± 6 Ma) |
| Age of structures? | ~3,200 Ma: Undoubted primary (bio)sedimentary structures |
| Geological context? | Found within a well preserved, little deformed sandstone of low metamorphic grade; interpreted tidal flat environment; restricted to the tops of decimetre-scale fining upwards sequences; viable for life |
| Biological morphology? | Wrinkle structures: ~5 mm wavelength and ~3 mm height; roll up structures; interwoven sandstone and carbon rich laminae (above right); quartz grains, with their long axes aligned, floating in a matrix of carbonaceous laminae |
| Biogeochemistry? | Microstructures are carbonaceous; bulk $\delta^{13}C$ values of –20.1‰ to –21.5‰ PDB |
| Type of organism? | Unknown photoautotrophic mat builder |
| INTERPRETATION IN THE LITERATURE | Microbially induced sedimentary structures (Noffke et al., 2006b) |
| **IS THIS REALLY LIFE?** | **A very promising type of biosignature. Experiments now needed to *ensure* structures cannot be replicated in a non-biological carbon-rich environment** |

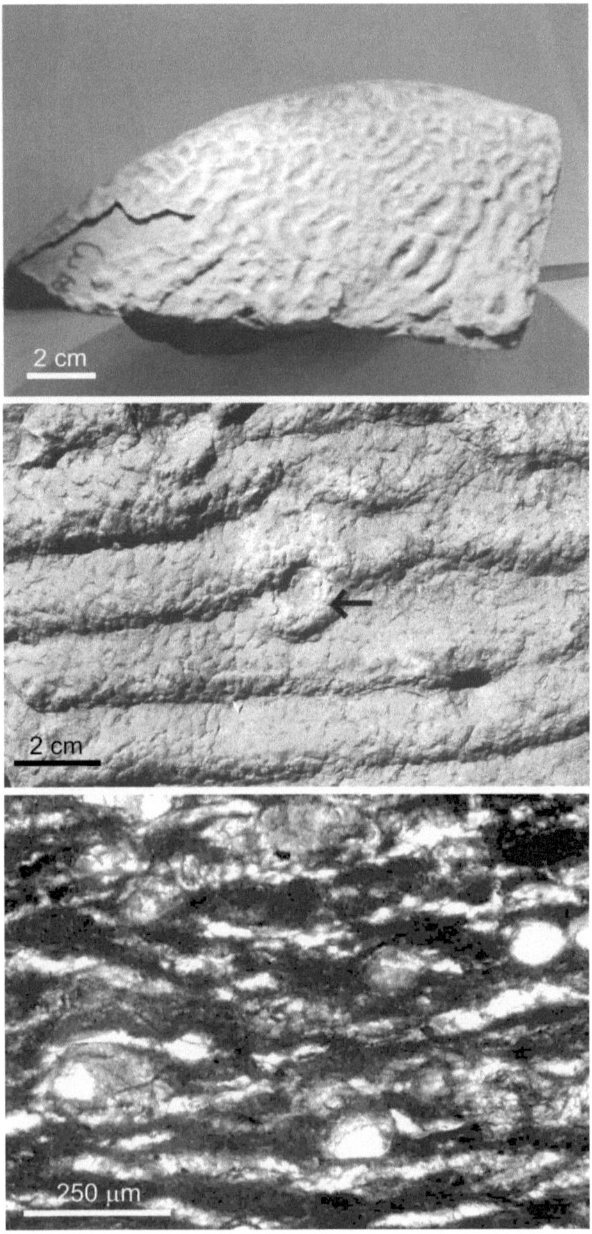


**Fig. B77** Younger analogues for the putative microbially influenced sedimentary structures described from the Moodies Group. *Top*, wrinkle structures representing a buried microbial mat from the ~2,950 Ma Brixton Formation, South Africa. *Middle*, wrinkle structures superimposed on sedimentary ripples, with possible trace fossil (*arrowed*), from the ~550 Ma Nama Group, Namibia. *Bottom*, modern siliciclastic microbial mat from Mellum Island, North Sea, showing meshwork fabric of quartz grains supported in a network of carbonaceous filaments and laminations (compare to Fig. B76) (Images courtesy of Nora Noffke)

the silica substrates, sometimes leaving behind features termed etch pits which can potentially reflect the exact morphology of the etching microorganism (Brehm et al., 2005, Fig. 3b). Because biological growth exhibits spatial heterogeneity, these features are often concentrated in natural depressions in the mineral grain, joins between crystal lattices, or where the lattices have high concentrations of defects. This may in turn lead to the widening of pores and cracks and ultimately the complete dissolution of host mineral grains.

The only traces of this activity likely to be preserved in the Archean rock record are the actual etch pits. Unfortunately, these are usually very simple rounded or oval depressions so that, on the rare occasions when they are preserved in the rock record, formation by biological means is very difficult to prove.

## Recommended Reading

Bennett, P., and Siegel, D. I., 1987, Increased solubility of quartz in water due to complexing by organic compounds, *Nature* **326**: 684–686.

Brehm, U., Gorbushna, A., and Mottershead, D., 2005, The role of microorganisms and biofilms in the breakdown and dissolution of quartz and glass, *Palaeogeography Palaeoclimatology Palaeoecology* **219**: 117–129.

Gerdes, G., Klenke, T., and Noffke, N., 2000, Microbial signatures in peritidal siliciclastic sediments: a catalogue, *Sedimentology* **47**: 279–308.

Noffke, N., 2000, Extensive microbial mats and their influences on the erosional and depositional dynamics of a siliciclastic cold water environment (lower Arenigian, Montagne Noire, France), *Sedimentary Geology* **136**: 207–215.

Noffke, N., Gerdes, G., Klenke, Th., and Krumbein, W. E., 2001, Microbially induced sedimentary structures—a new category within the classification of primary sedimentary structures, *Journal of Sedimentary Research* **71**: 649–656.

Noffke, N., Knoll, A. H., and Grotzinger, J., 2002, Sedimentary controls on the formation and preservation of microbial mats in siliciclastic deposits: a case study from the upper Neoproterozoic Nama Group, Namibia, *Palaios* **17**: 1–14.

Noffke, N., Hazen, R., and Nhleko, N., 2003, Earth's earliest microbial mats in a siliciclastic marine environment (2.9 Ga Mozaan Group, South Africa), *Geology* **31**: 673–676.

Noffke, N., Beukes, N., Gutzmer, J., and Hazen, R., 2006a, Spatial and temporal distribution of microbially induced sedimentary structures: a case study from siliciclastic storm deposits of the 2.9 Ga Witwatersrand Supergroup, South Africa, *Precambrian Research* **146**: 35–44.

Noffke, N., Eriksson, K., A., Hazen, R. M., and Simpson, E. L., 2006b, A new window into Early Archean life: microbial mats in Earth's oldest siliciclastic tidal deposits (3.2 Ga Moodies Group, South Africa), Geology 34: 253–256.

# Chapter 18
# ~3,200 Ma, Dixon Island Formation*, Cleaverville Greenstone Belt, West Pilbara, Western Australia

**Fig. B78** The ~350 m thick Dixon Island Formation comprises rhyolitic tuff, black chert (pictured *centre* of photograph) and other varicoloured chert interpreted to have formed in a low temperature hydrothermal setting some 500–2,000 m below the ocean surface. Putative microbial material is found in both massive and laminated black chert, in the form of spirals, rods, spheres and dendrites, together with carbonaceous grains and a stromatolite-like mat. Scale: person approximately 2 m tall, sequence youngs to the left (Photograph courtesy of Shoichi Kiyokawa)

---

*Some uncertainty exists over where exactly this Formation should be placed within the West Pilbara stratigraphy.

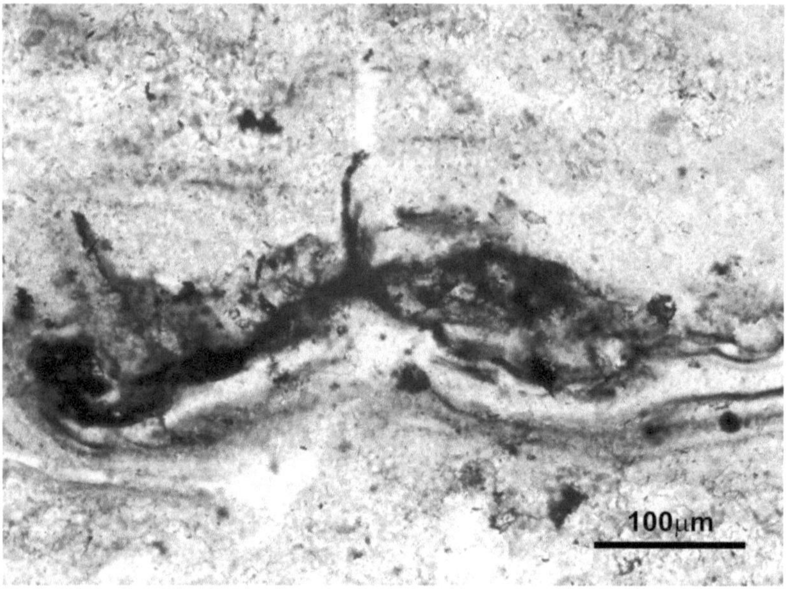


**Fig. B79** Thin section image of part of a putative microbial mat from the Dixon Island Formation (Image courtesy of S. Kiyokawa)

| | |
|---|---|
| Age of host rock? | ~3,200 Ma: Black chert, Dixon Island Formation, Western Australia; geochronology from rhyolitic tuff in the middle of the Formation, dated at 3,195 ± 12 Ma |
| Age of structures? | ~3,200 Ma: Preservation of the structures by silicification is interpreted to be syn-depositional, before eruption of the overlying non-silicified pillow basalts; structures occur away from fractures and veins |
| Geological context? | Surface of the ocean floor around a hydrothermal vent system – this context was determined by high resolution geological mapping and appears viable for life |
| Biological morphology? | Clumps and laminations of carbonaceous material with filaments rising sub-vertically (pictured above) |
| Biogeochemistry? | Carbonaceous; bulk $\delta^{13}C = -33‰$ to $-27‰$ PDB; Raman spectra consistent with known metamorphic grade |
| Type of organism? | Possibly iron oxidising anoxygenic bacteria flourishing around hydrothermal vents (Kiyokawa et al., 2006) |
| INTERPRETATION IN THE LITERATURE | Remnants of a microbial mat (Kiyokawa et al., 2006) |
| **IS THIS REALLY LIFE?** | **Possible life, but the morphology is not uniquely biogenic. *In situ* geochemistry now needed to support the thorough geological mapping by Kiyokawa et al.** |

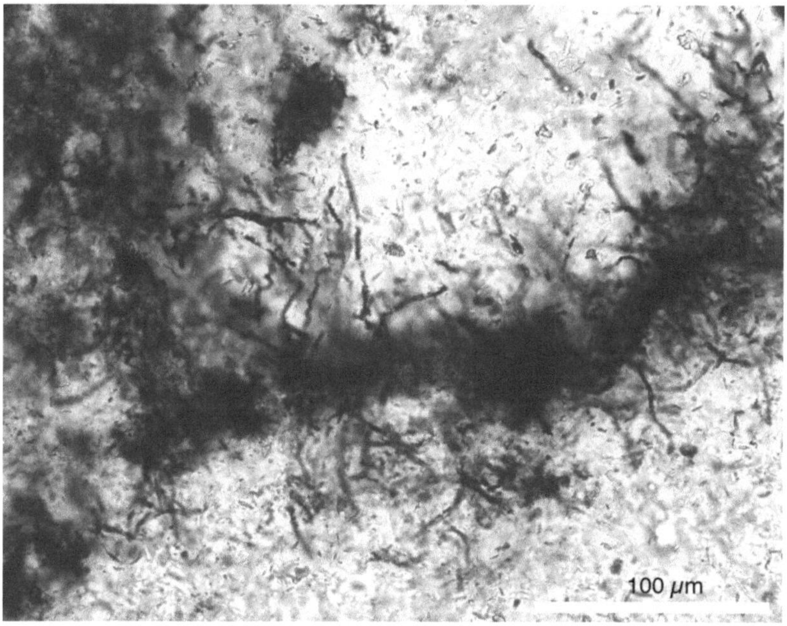


**Fig. B80** Putative filamentous microfossils from the Dixon Island Formation (Image courtesy of S. Kiyokawa)

| | |
|---|---|
| Age of host rock? | ~3,200 Ma: Black chert, Dixon Island Formation, Western Australia; geochronology from rhyolitic tuff in the middle of the Formation, dated at 3,195 ± 12 Ma |
| Age of structures? | ~3,200 Ma: Structures occur in a laterally continuous stratiform bedded chert; some are cross cut by later veins; Raman signature and H/C ratios suggest they are not modern contaminants |
| Geological context? | Ocean floor at an estimated 500–2,000 m palaeodepth in an island arc-type setting, around a hydrothermal vent system; viable for life |
| Biological morphology? | Filaments: 1 μm wide, ~100 μm long; associated with a carbonaceous laminated layer within the host chert |
| Biogeochemistry? | Carbonaceous; bulk $\delta^{13}C$ = −33‰ to −27‰ PDB; Raman spectra consistent with known metamorphic grade |
| Type of organism? | Possibly iron oxidising anoxygenic bacteria flourishing around hydrothermal vents (Kiyokawa et al., 2006) |
| INTERPRETATION IN THE LITERATURE | Possible microfossils (Kiyokawa et al., 2006) |
| **IS THIS REALLY LIFE?** | **Possible life. Much hinges on the interpreted temperature of the hydrothermal system as to whether non-biological carbonaceous artefacts may be a more likely explanation. *In situ* biogeochemistry also needed** |

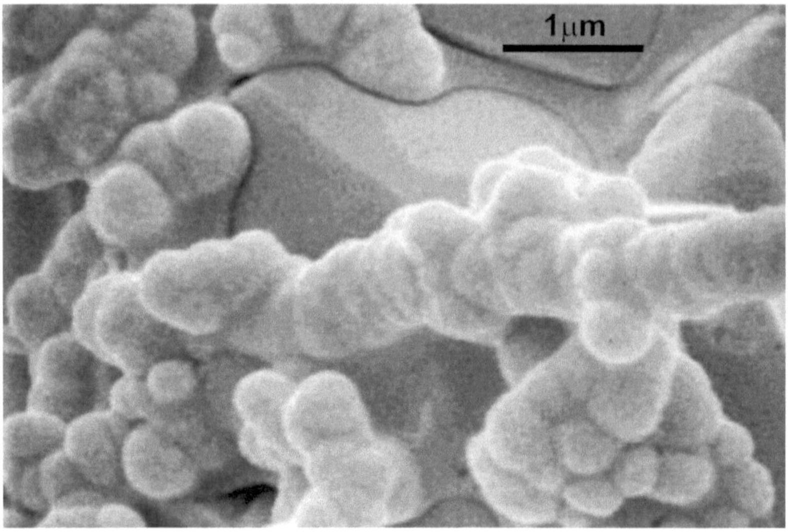


**Fig. B81** SEM image of possible spherical microfossils from the Dixon Island Formation (Image courtesy of S. Kiyokawa)

| | |
|---|---|
| Age of host rock? | ~3,200 Ma: Black chert, Dixon Island Formation, Western Australia; geochronology from rhyolitic tuff in the middle of the Formation, dated at 3,195 ± 12 Ma |
| Age of structures? | Preservation of the structures by silicification is *interpreted* to be syn-depositional, before eruption of the overlying non-silicified pillow basalts (but see below) |
| Geological context? | Ocean floor surface near to hydrothermal vents; viable for life |
| Biological Morphology? | Spheres: ~1 μm in diameter; sometimes linked in chains (pictured above) |
| Biogeochemistry? | Carbonaceous; bulk $\delta^{13}C = -33‰$ to $-27‰$; Raman spectra consistent with known metamorphic grade |
| Type of organism? | Unknown |
| INTERPRETATION IN THE LITERATURE | Microfossils (Kiyokawa et al., 2006) |
| **IS THIS REALLY LIFE?** | **Possible life, but found *between* quartz grains (not embedded) under SEM examination – could be non-indigenous** |

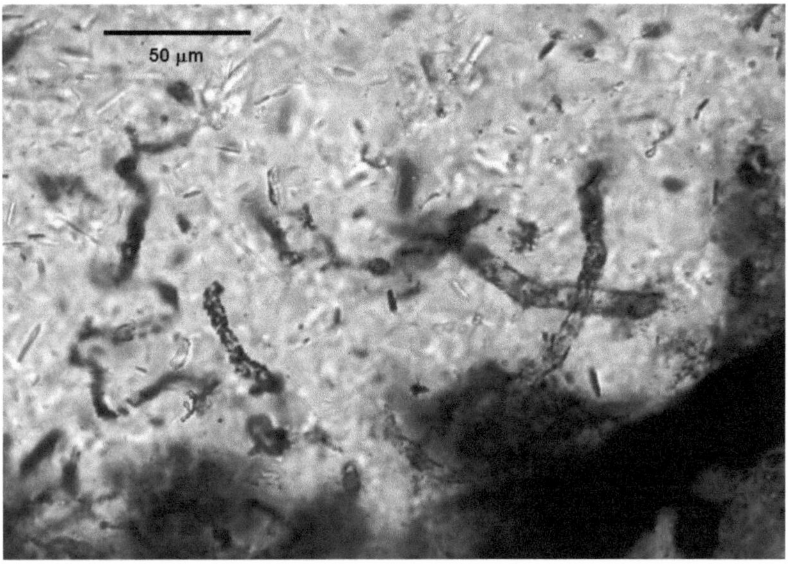


**Fig. B82** Putative spiral microfossils from the Dixon Island Formation (Image courtesy of Shoichi Kiyokawa)

| | |
|---|---|
| Age of host rock? | ~3,200 Ma: Black chert, Dixon Island Formation, Western Australia; geochronology from rhyolitic tuff in the middle of the Formation, dated at 3,195 ± 12 Ma |
| Age of structures? | ~3,200 Ma: Occur in laterally extensive bedded chert; petrographic relationships and micro-mapping suggests syn-depositional silicification and preservation; occur away from fractures and veins |
| Geological context? | Ocean floor surface near a hydrothermal vent system; viable for life |
| Biological morphology? | Spiral filaments: ~10 μm in diameter, 50–150 μm long; cylindrical, with <1 μm thick walls; small septa visible |
| Biogeochemistry? | Carbonaceous; bulk $\delta^{13}C = -33‰$ to $-27‰$; Raman spectra consistent with known metamorphic grade |
| Type of organism? | Unknown, but must be moderately thermophilic and capable of living at inferred 500–2,000 m depth |
| INTERPRETATION IN THE LITERATURE | Microfossil (Kiyokawa et al., 2006) |
| **IS THIS REALLY LIFE?** | **Possible life, but images do not clearly show all the described biological features (e.g. septa); structures could also be ambient inclusion trails (cf. Figs B15, B45, B46) with the terminal crystal out of the plane of view, although this does not preclude a biological explanation** |

## Recommended Reading

Hessler, R. R., and Kaharl, V.A., 1995, The deep-sea hydrothermal vent community: an overview. In: Humphris, S. E., Zierenberg, R. A., Mullineaux, L. S., and Thomsom, R. E. (Eds.) *Seafloor Hydrothermal Systems- Physical Chemical, Biological, and Geological Interactions, Geophysical Monograph* **91**: 72–84.

Kiyokawa, S., and Taira, A., 1998, The Cleaverville Group in the West Pilbara coastal granitoid-greenstone terrain of Western Australia: an example of a Middle Archean immature oceanic island-arc succession, *Precambrian Research* **88**: 109–142.

Kiyokawa, S., Taira, A., Byrne, T., Bowring, S., and Sano, Y., 2002, Structural evolution of the middle Archean coastal Pilbara terrane, Western Australia, *Tectonics* **21(5)**: 8–1–8–24.

Kiyokawa, S., Ito, T., Ikehara, M., and Kitajima, F., 2006, Middle Archean volcano-hydrothermal sequence: bacterial microfossil-like bearing 3.2 Ga Dixon Island Formation, coastal Pilbara terrane, Australia, *Bulletin of the Geological Society of America* **118**: 3–22.

# Chapter 19
# ~3,000 Ma, Cleaverville Formation, Cleaverville Greenstone Belt, West Pilbara, Western Australia

**Fig. B83** The Cleaverville Formation is composed of bedded chert, banded iron formation, siltstone, sandstone, and minor volcaniclastic rocks. Possible evidence for life has been found in a bedded chert unit near to Cleaverville Beach (pictured *above*). Scale is given by the dirt road running parallel to the beach (Photograph courtesy of the Shire of Roebourne Visitor Centre)

D. Wacey, *Early Life on Earth; A Practical Guide,*

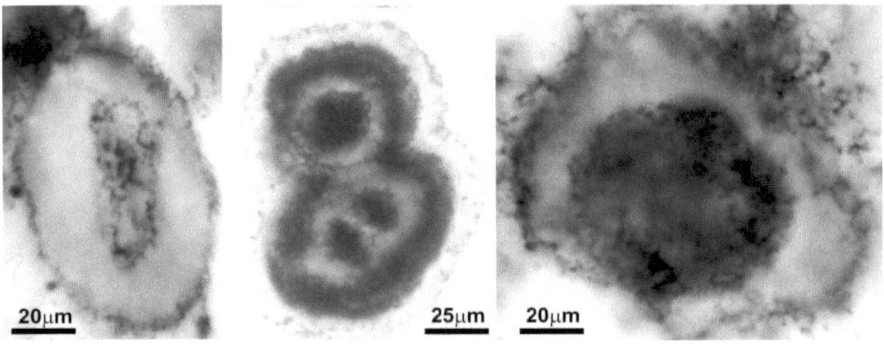


**Fig. B84** Regular and irregular carbonaceous spheroids from the Cleaverville Formation (Images courtesy of Yuichiro Ueno)

| | |
|---|---|
| Age of host rock? | ~3,000 Ma: Cleaverville Formation, Western Australia; geochronology from intrusive granophyre, dated at 3,014 ± 6 Ma and detrital zircons dated at 3,018 ± 3 Ma |
| Age of structures? | Varies: Structures occur in petrographic thin sections away from cracks and veins; some are compressed parallel to bedding and are interpreted as primary; others are demonstrably secondary diagenetic features |
| Geological context? | Occur in laterally continuous black chert with parallel lamination, originally deposited as carbonaceous mud – viable for life. However, depositional environment is controversial (has been suggested to be mid-ocean ridge, island arc or continental margin) |
| Biological morphology? | Spheroids; 4–89 μm in diameter; bimodal size distribution; solitary and paired, usually enclosed by outer envelopes; thin (<1 μm) walls; numerous (15,000 in seven thin sections); some elongate parallel to bedding; many have internal kerogenous structures; some are irregularly deformed indicative of post-mortem degradation |
| Biogeochemistry? | Structures are carbonaceous/kerogenous |
| Type of organism? | Resemble ensheathed, colonial, coccoid cyanobacteria |
| INTERPRETATION IN THE LITERATURE | Possible microfossils (the primary ones)<br>Inorganic diagenetic artefacts (the secondary ones) (Ueno et al., 2006) |
| **IS THIS REALLY LIFE?** | **Paired morphology does not necessarily indicate cell division. Most spheroids are regular and concentric which might be expected for inorganic precipitates from episodic crystal growth. Generally larger than typical prokaryote cells** |

## Recommended Reading

Hickman, A. H., Smithies, R. H., Pike, G., Farrell, T. R., and Beintema, K. A., 2001, Evolution of the West Pilbara granite-greenstone terrane and Mallina Basin, Western Australia—a field guide, *Western Australia Geological Survey Record* **2001/16**: 63p.

Kiyokawa, S., and Taira, A., 1998, The Cleaverville Group in the West Pilbara coastal granitoid-greenstone terrain of Western Australia: an example of a Mid-Archaean immature oceanic island-arc succession, *Precambrian Research* **88**: 109–142.

Ohta, H., Maruyama, S., Takahashi, E., and Kato, Y., 1996, Field occurrence, geochemistry, and petrogenesis of the Archean Mid-Oceanic Ridge Basalts (AMORBs) of the Cleaverville area, Pilbara craton, Western Australia, *Lithos* **37**: 199–221.

Ueno, Y., Isozaki, Y., and McNamara, K.J., 2006, Coccoid-like microstructures in a 3.0 Ga chert from Western Australia, *International Geology Review* **48**: 78–88.

# Chapter 20
# ~3,000 Ma, Farrel Quartzite, East Pilbara, Western Australia

**Fig. B85** The Farrel Quartzite is a mostly clastic sedimentary unit with thin inter-beds of black, and banded black-and-white, chert. Putative microfossils have been found within these chert horizons (pictured *above*) in the vicinity of Mt Grant and Mt Goldsworthy in Western Australia. The Farrel Quartzite is the lowermost member of the Gorge Creek Group which was deposited between about 3,190 Ma and 2,970 Ma. Scale: pen is ~15 cm long (Photograph courtesy of Ken Sugitani)

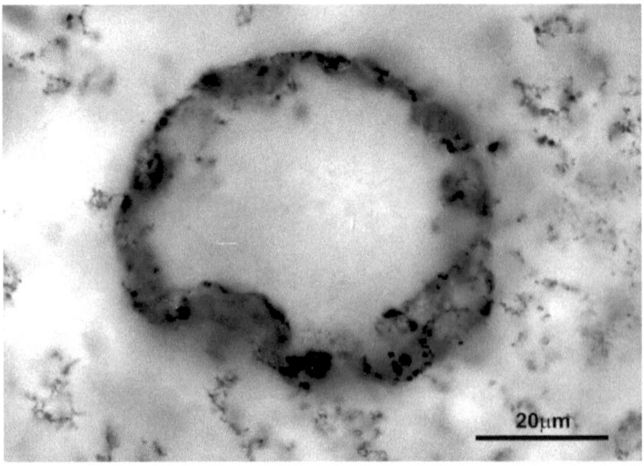


**Fig. B86** Possible spherical microfossil from the Farrel Quartzite (Image courtesy of Ken Sugitani)

| | |
|---|---|
| Age of host rock? | ~3,000 Ma: Farrel Quartzite, Pilbara Craton, Western Australia; age constrained by geochronology – 2,970 Ma on overlying unit and 3,190 Ma on underlying unit |
| Age of structures? | ~3,000 Ma: Antiquity demonstrated by petrographic relationships between the microfossil and its host spherulitic chert, and by cross cutting vein relationships |
| Geological context? | Occur in laterally traceable bedded cherts and in chert clasts within inter-bedded sandstone. Rocks are sedimentary, deposited in a shallow-water, continental margin setting. Viable for life |
| Biological morphology? | Spheres: 2.5–80 μm in diameter with bimodal distribution; mostly hollow with ~1 μm thick uneven walls; walls can often be broken or distorted; occur on their own and, more rarely, as aggregates; sometimes co-occur with carbonaceous films and sheets |
| Biogeochemistry? | Laser Raman shows structures are carbonaceous and of similar metamorphic grade to other carbon in the matrix; bulk $\delta^{13}C$ = −31.9‰ to −35.4‰ PDB |
| Type of organism? | Unknown |
| INTERPRETATION IN THE LITERATURE | Probable microfossils (Sugitani et al., 2007) |
| **IS THIS REALLY LIFE?** | **Possible life, but simple spheroidal morphologies are a common non-biological crystallization artefact** |

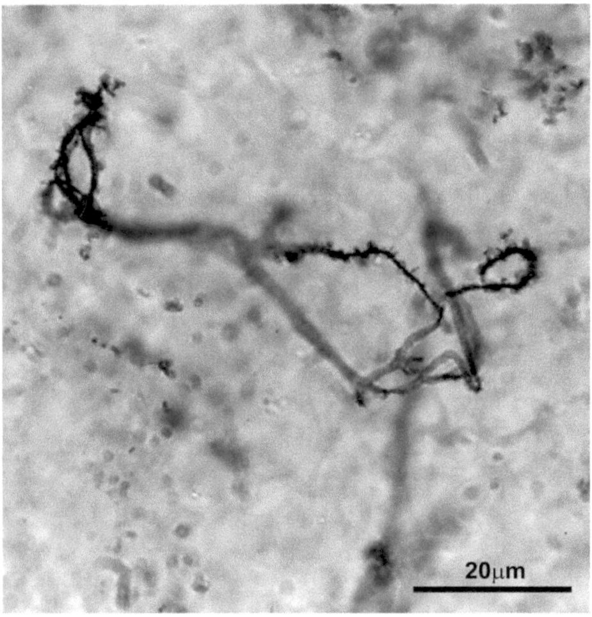


**Fig. B87** Putative thread-like microfossil from the Farrel Quartzite (Image courtesy of K. Sugitani)

| | |
|---|---|
| Age of host rock? | ~3,000 Ma: Farrel Quartzite, Pilbara Craton, Western Australia; age constrained by geochronology – 2,970 Ma on overlying unit and 3,190 Ma on underlying unit |
| Age of structures? | Occur in secondary cavity fill or vein chert so could be much younger than 3,000 Ma |
| Geological context? | Host rocks are sedimentary, deposited in continental margin, shallow-water setting, but context of threads is poorly constrained |
| Biological morphology? | Thread-like: <1 μm in diameter, 10's–100's μm in length; often twisted, occasionally branched; solid rather than hollow; too simple to confidently attribute to biology |
| Biogeochemistry? | Laser Raman shows structures are carbonaceous and of similar metamorphic grade to matrix carbon; bulk $\delta^{13}C$ = −31.9 to −35.4 PDB |
| Type of organism? | Unknown |
| INTERPRETATION IN THE LITERATURE | Dubiofossil (Sugitani et al., 2007) |
| **IS THIS REALLY LIFE?** | **Unlikely. Both the geological context and biogenicity are uncertain. Bulk $\delta^{13}C$ may suffer from contamination from other structures** |

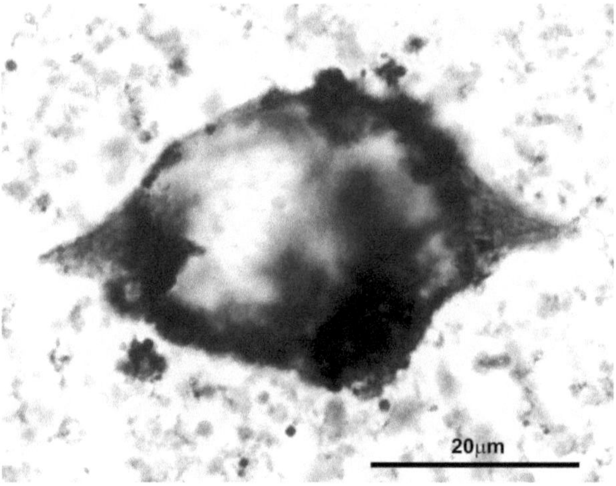


**Fig. B88** Putative lens-like microfossil from the Farrel Quartzite (Image courtesy of Ken Sugitani)

| | |
|---|---|
| Age of host rock? | ~3,000 Ma: Farrel Quartzite, Pilbara Craton, Western Australia; age constrained by geochronology – 2,970 Ma on overlying unit and 3,190 Ma on underlying unit |
| Age of structures? | ~3,000 Ma: Antiquity demonstrated by fabric mapping of the host spherulitic chert, and by cross cutting vein relationships |
| Geological context? | Occur in laterally traceable bedded cherts and in chert clasts within inter-bedded sandstone; these are demonstrably sedimentary, deposited in a shallow-water setting and are therefore viable for life |
| Biological morphology? | Lens or spindle shaped: 20–40 μm in length, 15–35 μm in width; main body is hollow with wall of up to 5 μm; appendages are 2–10 μm in length, tapering and solid; usually solitary but can occur as pairs, chains or aggregates; have occasional internal spheroidal bodies |
| Biogeochemistry? | Laser Raman shows structures are carbonaceous and of similar metamorphic grade to carbon elsewhere in the matrix; bulk $\delta^{13}C = -31.9$ to $-35.4$ PDB |
| Type of organism? | Unknown |
| INTERPRETATION IN THE LITERATURE | Probable microfossil (Sugitani et al., 2007) |
| **IS THIS REALLY LIFE?** | **Possible life, *in situ* carbon isotopes and elemental analysis now needed. Problematical morphology with no obvious modern analogue** |

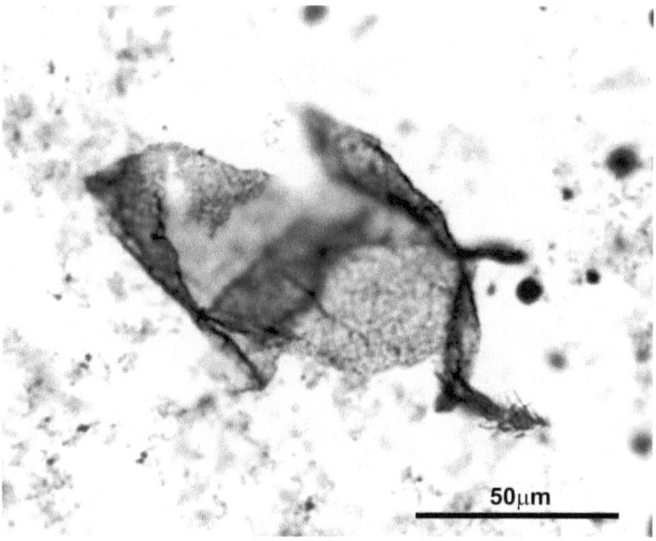


**Fig. B89** Possible biofilm from the Farrel Quartzite (Image courtesy of Ken Sugitani)

| | |
|---|---|
| Age of host rock? | ~3,000 Ma: Farrel Quartzite, Pilbara Craton, Western Australia; age constrained by geochronology of 2,970 Ma on overlying unit and 3,190 Ma on underlying unit |
| Age of structures? | ~3,000 Ma: Antiquity demonstrated by petrographic relationships between the microstructure and the host chert, and by cross cutting vein relationships |
| Geological context? | Occur in laterally traceable bedded cherts and in chert clasts within interbedded sandstone; host unit is sedimentary, deposited in shallow-water setting and is viable for life |
| Biological morphology? | Crumpled, folded and twisted sheets: 50–500 μm wide and 1–5 μm thick; multi-layer sheets are common; edges are thickened (possibly rolled up pictured above) and commonly notched or torn; can be associated with semi-hollow spheres; show stages of taphonomic degradation |
| Biogeochemistry? | Laser Raman shows structures are carbonaceous and of similar metamorphic grade to matrix carbon; bulk $\delta^{13}C = -31.9‰$ to $-35.4‰$ PDB |
| Type of organism? | Film (mat) building microbes |
| INTERPRETATION IN THE LITERATURE | Probable biofilm (Sugitani et al., 2007) |
| **IS THIS REALLY LIFE?** | **Context, morphology and taphonomic degradation all consistent with life. *In situ* isotopic and elemental analysis now needed to strengthen claim further** |

## Recommended Reading

Smithies, R. H., Van Kranendonk, M. J., and Hickman, A. H., De Grey, W.A., 2004, Sheet 2757 (Version 2.0), *Western Australia Geological Survey* 1:100 000 Geological Series, Perth.

Sugitani, K., Yamamoto, K., Adachi, M., Kawabe, I., and Sugisakli, R., 1998, Archean cherts derived from chemical, biogenic and clastic sedimentation in a shallow restricted basin: examples from the Gorge Creek Group in the Pilbara Block, *Sedimentology* **45**: 1045–1062.

Sugitani, K., Mimura, K., Suzuki, K. Nagamine, K. and Sugisaki, R., 2003, Stratigraphy and sedimentary petrology of an Archaean volcanic-sedimentary succession at Mt Goldsworthy in the Pilbara block, Western Australia: implications of evaporite (Nahcolite) and barite deposition, *Precambrian Research* **120**: 55–79.

Sugitani, K., Yamashita, F., Nagaoka, T., Yamamoto, K., Minami, M., Mimura, K., and Suzuki, K., 2006a, Geochemistry and sedimentary petrology of Archean clastic sedimentary rocks at Mt. Goldsworthy, Pilbara Craton, Western Australia: evidence for the early evolution of continental crust and hydrothermal alteration, *Precambrian Research* **147**: 124–147.

Sugitani, K., Yamashita, F., Nagaoka, T., Minami, M., and Yamamoto, K., 2006b, Geochemistry of heavily altered Archean volcanic and volcaniclastic rocks of the Warrawoona Group, at Mt. Goldsworthy in the Pilbara Craton, Western Australia: implications for alteration and origin, *Geochemical Journal* **40**: 523–535.

Sugitani, K., Grey, K., Allwood, A., Nagaoka, T., Mimura, K., Minami, M., Marshall, C. P., Van Kranendonk, M. J., and Walter, M. R., 2007, Diverse microstructures from Archaean chert from the Mount Goldsworthy-Mount Grant area, Pilbara Craton, Western Australia: microfossils, dubiofossils, or pseudofossils? *Precambrian Research* **158**: 228–262.

# Chapter 21
# THE IMPOSTERS: Younger Biological Contaminants and Non-Biological Artefacts

This section contains images of the various types of contaminants and artefacts that may be confused for ancient life in early Archean rocks. Contaminants may be introduced by hydrothermal and metamorphic fluids at various stages throughout the (often complex) history of the host rock unit, or they may be introduced by weathering whenever the rock unit is exposed at or near the surface of the Earth. Contaminants may be very young indeed, introduced by modern soils or microbes, or even from human activity during sample collection, processing and preparation.

Various inorganic structures can also be confused for ancient life. In the main, these are purely chemical precipitates that can be misinterpreted as microfossils and/or biological stromatolites. Some types of physical sedimentary (e.g., ripples) and deformational structures (e.g., micro-folds) can also be mistaken for biological stromatolites, whilst non-biological processes can mimic biological isotope signatures (see also Fig. A35).

D. Wacey, *Early Life on Earth; A Practical Guide*,

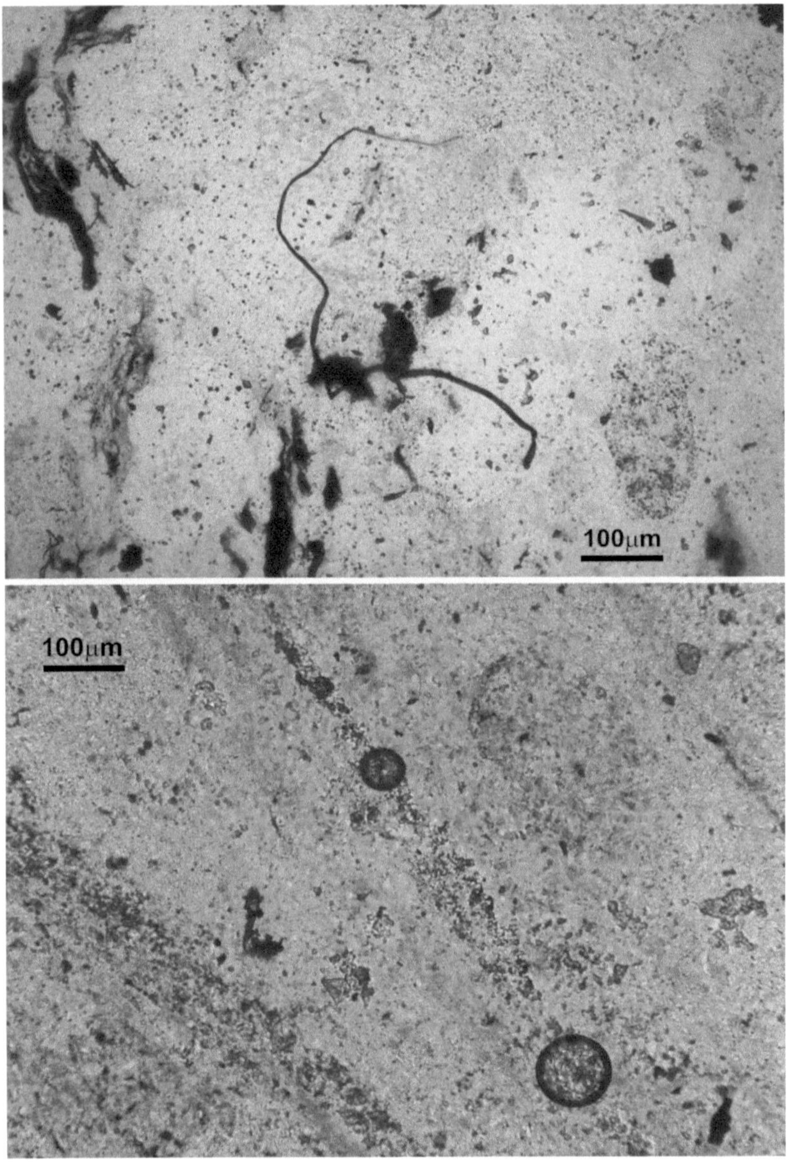


**Fig. B90** Preparation artefacts. *Top*, hair or clothes fibre on the surface of a thin section. This may be mistaken for a filamentous microfossil but can be distinguished by the fact that it cross cuts all elements of the thin section and has a slightly different focal depth to the rest of the thin section. *Bottom*, spherical bubbles of thin section mounting media. These may be mistaken for coccoid microfossils but can be distinguished by their very high relief, large size variation and different depth of focus to the remainder of the thin section (Images are from the 1,900 Ma Gunflint chert housed in the Oxford Palaeobiology Collections)

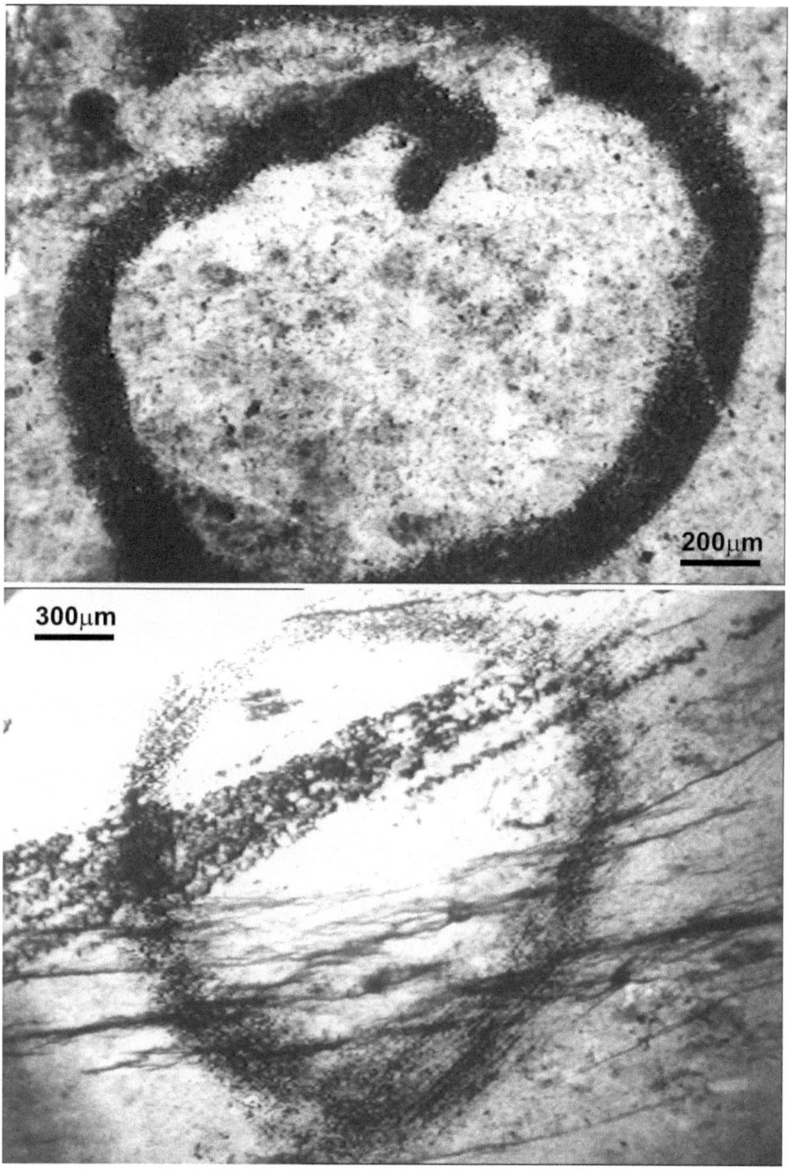


**Fig. B91** Man made artefacts. *Top*, marker pen artefact. *Bottom*, pencil artefact. These are generally an order of magnitude larger than microfossils and will be visible with the naked eye on the surface of the thin section. These are only really problematical when looking at second hand or loaned material. These should be avoided completely by using stage co-ordinates to record areas of interest within the thin section. Images are from the 3,460 Ma Apex chert, Western Australia (Samples are courtesy of Martin Brasier and the Natural History Museum, London)

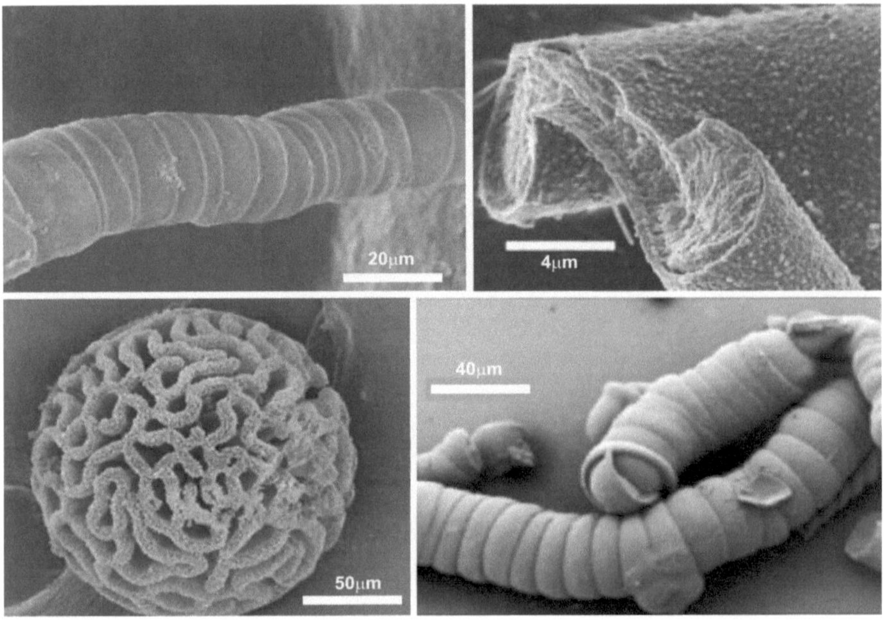


**Fig. B92** Complex non-biological artefacts – silica and barium carbonate (witherite) precipitates from laboratory experiments. Various filamentous 'abiomorphs' are produced that closely resemble filamentous microfossils. Larger complex floral spherulites (*lower left*) resembling brain coral have also been produced (*Top right* and *bottom right* from García-Ruiz et al., 2003; *top left* and *bottom left* from Hyde et al., 2004) (Images reproduced with permission from AAAS and Elsevier)

**Fig. B93** Filamentous non-biological structures. *Top*, xonotlite (hydrated calcium silicate) fibres showing close resemblance to silicified extra-cellular polymeric substances (EPS), or even some filamentous microfossils. *Bottom*, xonotlite whiskers, which could be mistaken for networks of silicified filamentous microbes. Both of these microstructures were formed non-biologically by mixing calcium hydroxide with silica gel or powder at ~250°C (Yanagisawa et al., 1997, reproduced with permission from Springer)

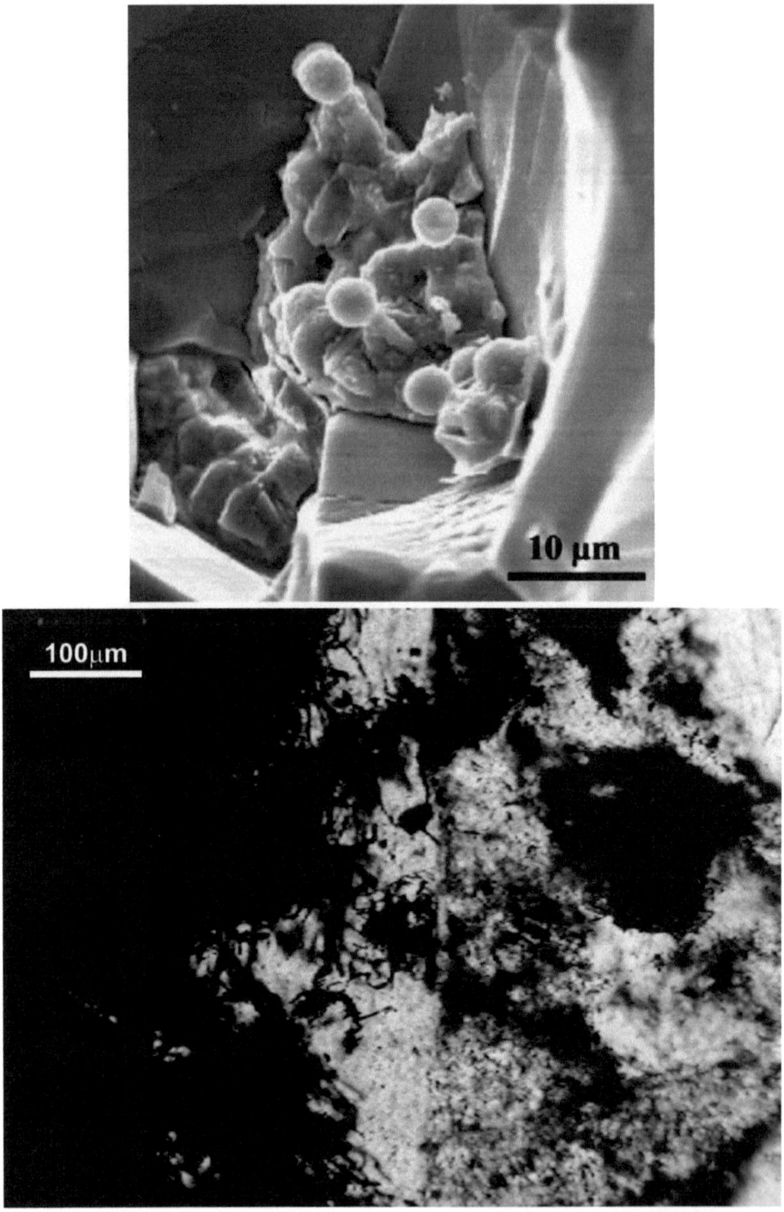


**Fig. B94** Modern endolithic contaminants. *Top*, coccoidal endoliths embedded in extra-cellular polymeric substances (EPS), found *in between* grains of quartz in a >3,700 Ma chert from the Isua supracrustal belt of south-west Greenland. These are interpreted to be no more than 8,000 years old (Westall and Folk, 2003, reproduced with permission from Elsevier). *Bottom*, modern endolithic microbes (fungal hyphae) invading a weathered margin of the ~3,400 Ma Strelley Pool sandstone from Western Australia (Image is from the Oxford Palaeobiology Collections)

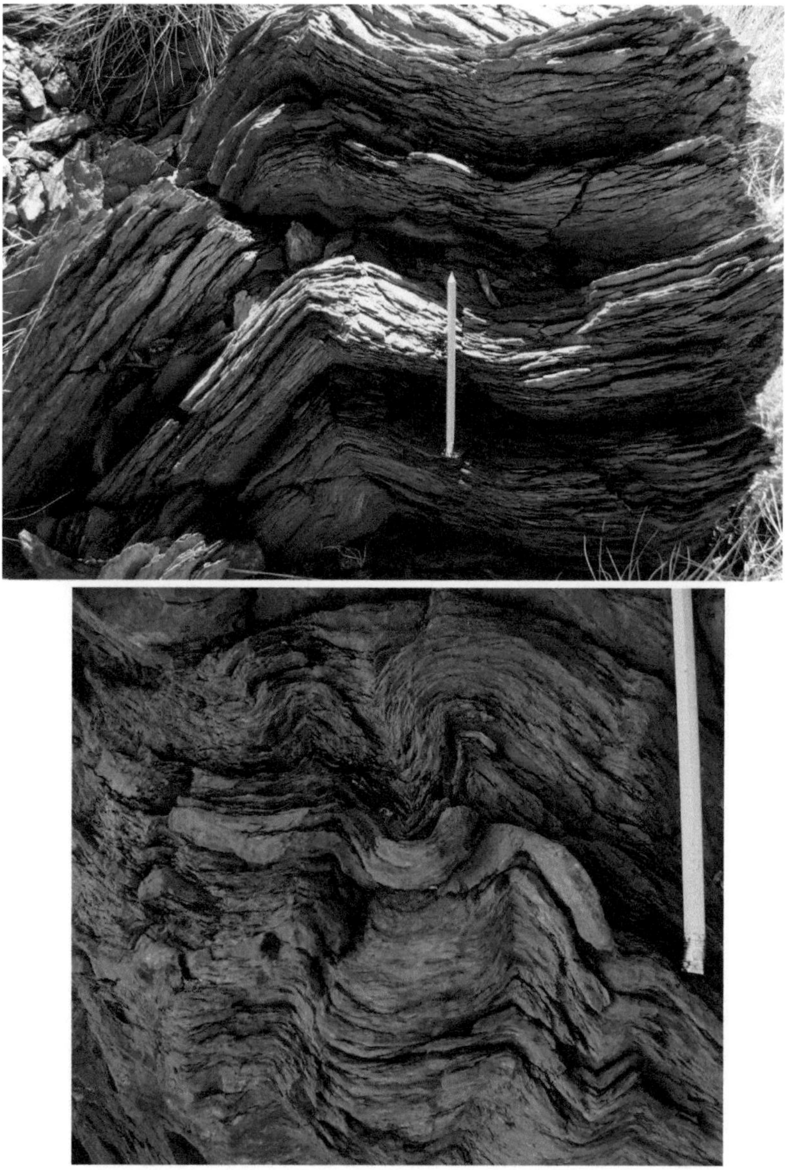

**Fig. B95** Non-biological stromatolite-like structures. *Top*, steep-sided tectonic fabric resembling coniform stromatolites. *Bottom*, s-shaped and z-shaped tectonic fabric resembling domed and branched stromatolites. Photographs taken by the author from a silicified volcanic unit within the ~3,450 Ma Panorama Formation, Western Australia. Scale: pencil is 18 cm long

**Fig. B96** Non-biological stromatolite-like structures. *Top*, the 'Taylor Stromatolite' in cross-section; this contains hundreds of ~50 μm thick multicoloured, convex up laminae that form laterally linked domes and columns with homogenised, un-laminated brown paint in some of the inter-column areas. It was formed as an overspray deposit that accumulated on a paint spray booth from an automobile production line. *Bottom*, columnar stromatolite-like structures created by sub-aerial deposition of titanium based colloids (paint), encased in resin and shown in vertical cross-section; the columns are inclined towards the 'sediment source' on the left and show bridging laminae between the columns (Samples are from the Oxford Palaeobiology Collections, and images are courtesy of Nicola McLoughlin)

**Fig. B97** Non-biological stromatolite-like structures. *Top*, plan view of linguoid ripples (a physical sedimentary structure) in the ~3,400 Ma Strelley Pool chert, Western Australia. Small amounts of deformation and/or erosion of these ripples may produce coniform structures (*bottom*) similar to those interpreted as stromatolites by Hoffman et al. (1999) and Allwood et al. (2006) (see pages 184 and 190–191) in the same rock unit. Scale: pen is 12 cm long (Images are courtesy of Martin Brasier)

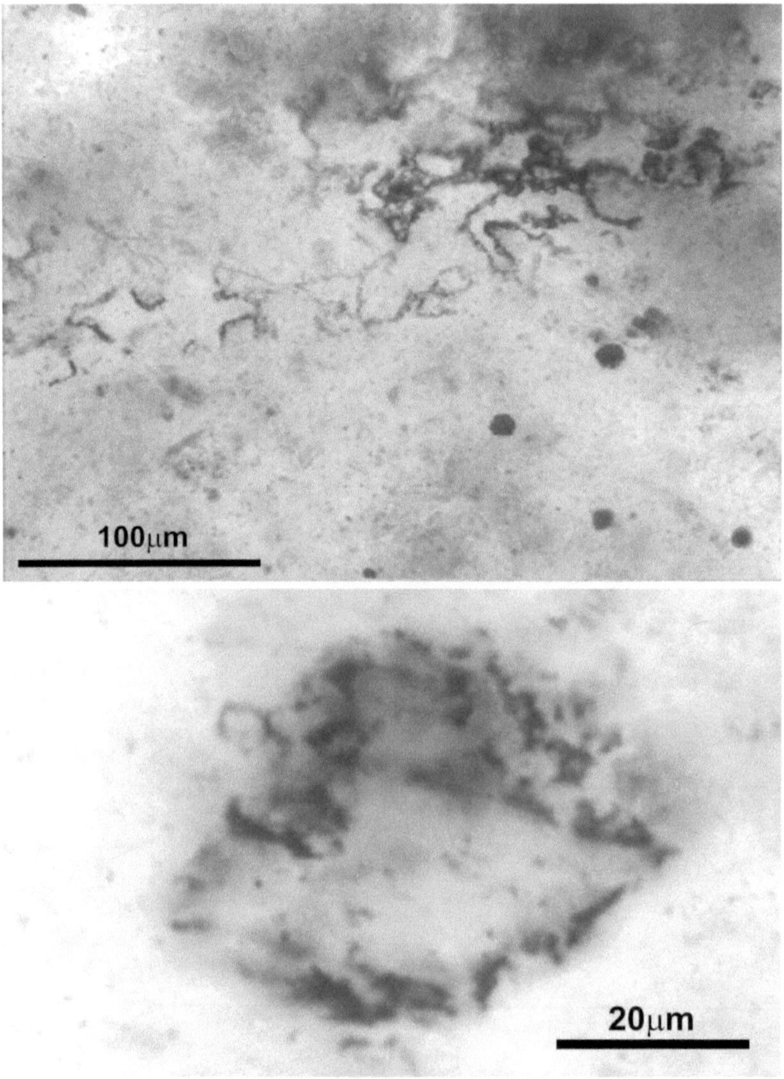


**Fig. B98** Crystal margin artefacts. *Top*, carbonaceous filaments forming part of a continuous, ramifying network (across *centre* of image), wrapped around angular crystal margins. Some areas of these, when viewed in isolation, have been mistaken for filamentous microfossils. *Bottom*, crystal rhomb (probably dolomite) coated by a carbonaceous film. Taken out of context, an isolated area of this coating may be mistaken for certain types of filamentous microfossils (Images are from the ~3,460 Ma Apex chert, Western Australia, courtesy of Martin Brasier)

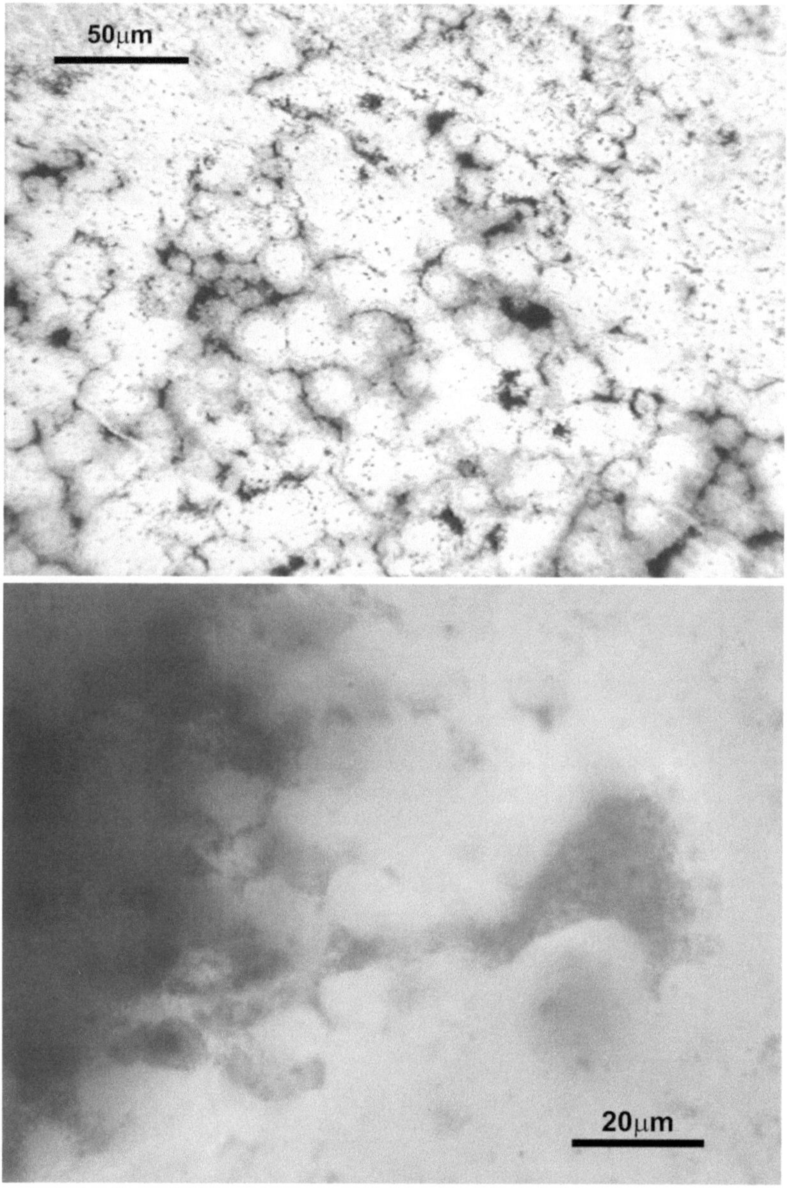


**Fig. B99** Spherulitic crystal margin artefacts. *Top*, carbonaceous crystal rims around spherulitic chalcedonic quartz from the 580 Ma Gwna Group of Wales. *Bottom*, similar spherulitic fabric from the ~3,460 Ma Apex chert, Western Australia. Parts of this type of fabric have previously been misinterpreted as septate filamentous microfossils (Images are courtesy of Martin Brasier)

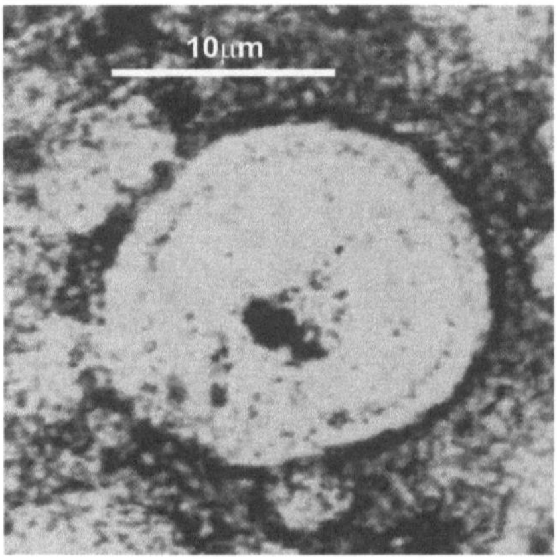


**Fig. B100** Spherical non-biological artefacts. An inorganic vesicular structure from a ~3,400 Ma pillow basalt (Engel et al., 1968; reproduced with permission of AAAS). The size and spheroidal microstructure, with a pseudo-'double wall' and central body, may lead to misidentification as a coccoid microfossil

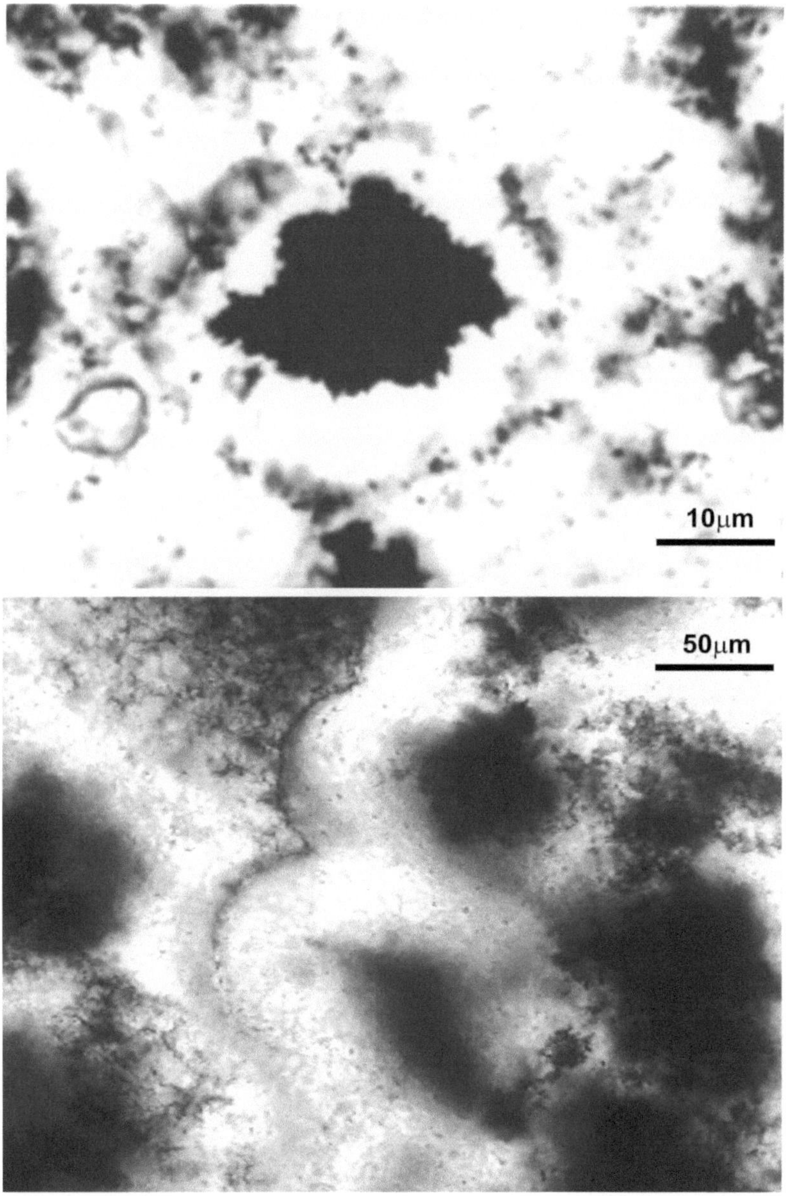


**Fig. B101** Sub-spherical non-biological artefacts. *Top*, 'Stellate haloed' object (Buick, 1990; reproduced under the fair usage policy of SEPM) interpreted to have formed during growth of chalcedonic silica spherulites. *Bottom*, botryoids of chalcedony growing from dark, fluffy carbonaceous material into a void space. Carbonaceous reaction rims occur around the botryoid margins (Image is from the ~3,460 Ma Apex chert, Western Australia, courtesy of Martin Brasier)

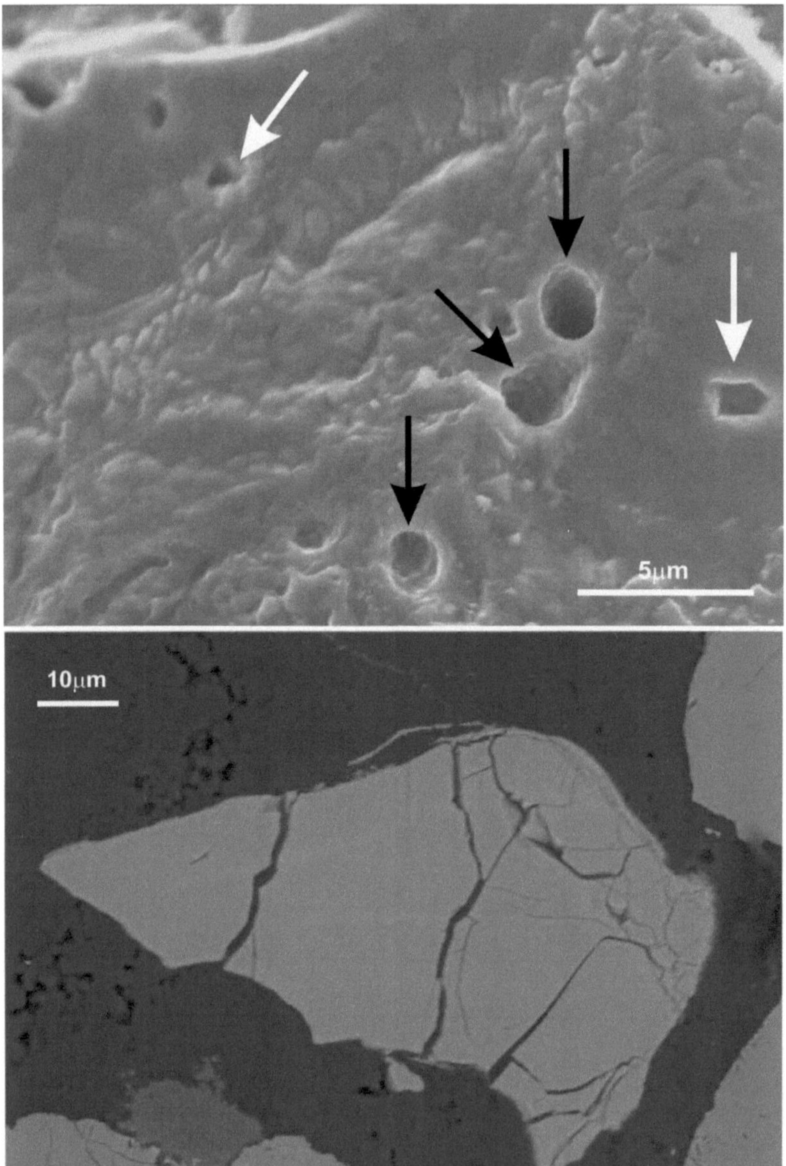


**Fig. B102** Physical artefacts. *Top*, surface of a pyrite grain showing putative microbial borings that are roughly spherical in cross section (*black arrows*) and non-biological angular erosion pits (*white arrows*). *Bottom*, a fractured pyrite grain. The angular nature, jigsaw-puzzle fit and changes in diameter of the fractures can be used to distinguish such features from microbial borings (Images are from the ~3,400 Ma Strelley Pool sandstone, Western Australia housed in the Oxford Palaeobiology Collections)

# Recommended Reading

Allwood, A. C., Walter, M. R., Kamber, B. S., Marshall, C. P., and Burch, I. W., 2006, Stromatolite reef from the Early Archaean era of Australia, *Nature* **441**: 714–718.

Awramik, S. M., Schopf, J. W., and Walter M. R., 1983, Filamentous fossil bacteria from the Archaean of Western Australia, *Precambrian Research* **20**: 357–374.

Brasier, M. D., Green, O. R., Jephcoat, A. P., Kleppe, A. K., Van Kranendonk, M. J., Lindsay, J. F., Steele, A., and Grassineau N. V., 2002, Questioning the evidence for Earth's oldest fossils, *Nature* **416**: 76–81.

Brasier, M. D., Green, O. R., Lindsay, J. F., and Steele, A., 2004, Earth's oldest (~3.5 Ga) fossils and the 'Early Eden' hypothesis: questioning the evidence, *Origins of Life and Evolution of the Biosphere* **34**: 257–269.

Brasier, M. D., Green, O. R., Lindsay, J. F., McLoughlin, N., Steele, A., and Stoakes, C., 2005, Critical testing of Earth's oldest putative fossil assemblage from the ~3.5 Ga Apex Chert, Chinaman Creek Western Australia, *Precambrian Research* **140**, 55–102.

Brasier, M. D., McLoughlin, N., and Wacey, D., 2006, A fresh look at the fossil evidence for early Archaean cellular life, *Philosophical Transactions of the Royal Society B* **361**: 887–902.

Brasier, M. D., Green, O. R., Lindsay, J. F., McLoughlin, N., Stoakes, C. A., Brasier, A., and Wacey, D., 2008, Earth's oldest putative fossil assemblage from the ~3.5Ga Apex chert, Chinaman Creek, Western Australia: a field and petrographic guide, *Geological Survey of Western Australia Record*

Buick, R., 1990, Microfossil recognition in Archaean rocks: an appraisal of spheroids and filaments from a 3500 MY old chert-barite unit at North Pole, Western Australia, *Palaios* **5**: 441–459.

Engel, A. E. J., Nagy, B., Nagy, L. A., Engel, C. G., Kremp, G. O. W., and Drew, C. M., 1968, Algal-like forms in Onverwacht Series, South Africa: oldest recognised lifelike forms on Earth, *Science* **161**: 1005–1008.

García-Ruiz, J. M., Hyde, S. T., Carnerup, A. M., Christy, A. G., Van Kranendonk, M. J., and Welham, N. J., 2003, Self-assembled silica carbonate structures and detection of ancient microfossils, *Science* **302**: 1194–1197.

Grotzinger, J. P., and Rothman, D. H., 1996, An abiotic model for stromatolite morphogenesis, *Nature* **383**: 423–425.

Hofmann, H. J., Grey, K., Hickman, A. H., and Thorpe, R.I., 1999, Origin of 3.45 Ga Coniform Stromatolites in the Warrawoona Group, Western Australia, *Bulletin of the Geological Society of America* **111**: 1256–1262.

Holm, N. G., and Charlou, J. L., 2001, Initial indications of abiotic formation of hydrocarbons in the Rainbow ultramafic hydrothermal system, Mid-Atlantic Ridge, *Earth and Planetary Science Letters* **191**: 1–8.

Horita, J., and Berndt, M. E., 1999, Abiogenic methane formation and isotopic fractionation under hydrothermal conditions, *Science* **285**: 1055–1057.

Hyde, S. T., Carnerup, A. M., Larsson, A. K., Christy, A. G., and Garcia-Ruiz, J. M., 2004, Self-assembly of carbonate-silica colloids: between living and non-living form, *Physica* **339**: 24–33.

Lowe, D. R., Byerly, G. R., Kyte, F. T., Shukloyokov, A., Asaro, F., and Krull, A., 2003, Early Archean spherule beds 3.47–3.34 Ga old in the Barberton greenstone belt, South Africa: a record of large meteor impacts and their influence on early crustal and biological evolution, *Astrobiology* **3**: 7–48.

McCollum, T. M., and Seewald, J. S., 2006, Carbon isotope composition of organic compounds produced by abiotic synthesis under hydrothermal conditions, *Earth and Planetary Science Letters* **243**: 74–84.

Oehler, J. H., 1976, Hydrothermal crystallization of silica gel, *Bulletin of the Geological Society of America* **87**: 1143–1152.

Sherwood Lollar, B., Westagate, T. D., Ward, J. A., Slater, G. F., and Lacrampe-Couloume, G., 2002, Abiogenic formation of alkanes in the Earth's crust as a minor source for global hydrocarbon reservoirs, *Nature* **416**: 522–524.

Simonson, B. M., 2003, Petrographic criteria for recognizing certain types of impact spherules in well preserved Precambrian successions, *Astrobiology* **3**: 49–64.

Westall, F., and Folk, R. L., 2003, Exogenous carbonaceous microstructures in Early Archaean cherts and BIFs from the Isua Greenstone Belt: implications for the search for life in ancient rocks, *Precambrian Research* **126**: 313–330.

Yanagisawa, K., Feng, Q., and Yamasaki, N., 1997, Hydrothermal synthesis of xonotlite whiskers by ion diffusion, *Journal of Materials Science Letters* **16**: 889–891.

# Index

**X**



**Z**

MIX
Papier aus verantwortungsvollen Quellen
Paper from responsible sources
FSC® C105338

FSC
www.fsc.org

If you have any concerns about our products,
you can contact us on
**ProductSafety@springernature.com**

In case Publisher is established outside the EU,
the EU authorized representative is:
**Springer Nature Customer Service Center GmbH**
**Europaplatz 3, 69115 Heidelberg, Germany**

Printed by Libri Plureos GmbH
in Hamburg, Germany